Implementing Event-Driven Microservices Architecture in .NET 7

Develop event-based distributed apps that can scale with ever-changing business demands using C# 11 and .NET 7

Joshua Garverick

Omar Dean McIver

BIRMINGHAM—MUMBAI

Implementing Event-Driven Microservices Architecture in .NET 7

Associate Group Product Manager: Gebin George

Publishing Product Manager: Kunal Sawant

Content Development Editor: Rosal Colaco

Technical Editor: Maran Fernandes

Copy Editor: Safis Editing

Project Coordinator: Manisha Singh

Proofreader: Safis Editing

Indexer: Subalakshmi Govindhan

Production Designer: Shyam Sundar Korumilli

Business Development Executive: Debadrita Chatterjee

Developer Relations Marketing Executive: Rayyan Khan and Deepak Kumar

First published: February 2023

Production reference: 2090323

Published by Packt Publishing Ltd.
Livery Place
35 Livery Street
Birmingham
B3 2PB, UK.

ISBN 978-1-80323-278-2

www.packtpub.com

To my wife, Melissa, and daughter, Audrey, for bringing me inspiration and supporting me while I go off and do things like write books.

– Josh Garverick

To my wife, Raegan, your love, support, and patience have made it possible for me to pursue my passion. And to my children, Olivia, Maxwell, and Benjamin, who bring light to my life with every new experience we share.

– Omar McIver

Foreword

In today's rapidly changing technology landscape, it's more important than ever to build software that can adapt to change and scale to meet the demands of users. Microservices have emerged as a popular architectural pattern for building flexible and scalable software systems. However, managing the complexity of microservice-based systems can be a daunting task.

Event-Driven Architecture (EDA) has become an increasingly popular solution to this challenge, as it allows services to communicate with each other in a loosely coupled manner, making it easier to build scalable and fault-tolerant systems. In this book, Josh Garverick and Omar McIver provide a comprehensive guide to implementing event-driven microservice architectures using .NET 7.

I have known Josh for a long time, and he has a wealth of experience in designing and building software systems using .NET. In this book, he shares his knowledge and expertise to help readers build better, more scalable, and more resilient microservice-based systems. He covers all the key concepts and techniques involved in implementing event-driven microservice architectures, including **Domain-Driven Design** (DDD), containerization, and implementing event-driven communication patterns.

What I particularly appreciate about this book is the practical, hands-on approach that Josh and Omar take. They provide an example application that illustrates how to apply EDA to real-world scenarios. They also cover the latest features of .NET 7, including minimal APIs, async streaming, and hot reload compatibility.

Having coined the phrase "*Rub a little DevOps on it*," I was thrilled to read the chapter on CI/CD Pipelines and Integrated Testing. Josh helps developers through the entire lifecycle of a microservices application including testing, observability, and even fault injection, and chaos testing.

This book is an essential resource for anyone involved in building microservice-based systems using .NET 7. Whether you're a seasoned developer or just getting started with microservices, Josh's and Omar's insights and expertise will help you build better, more scalable, and more resilient software systems. I highly recommend this book to anyone who wants to stay ahead of the curve in modern software development.

Donovan Brown

Partner Program Manager, Azure Incubations office of CTO, Microsoft Corporation

Contributors

About the authors

Josh Garverick, an eight-time Microsoft MVP in Azure, is the author of the book *Migrating to Azure*, a contributing author to *The Developer's Guide to Azure (2021)*, and the author of several online courses. Josh is a seasoned IT professional with more than 15 years of enterprise experience. He has worked in several large industry verticals, such as finance, healthcare, transportation, logistics, retail/consumer products, oil and gas, and insurance. He specializes in digital transformation as well as application life cycle management and is currently involved in enterprise cloud adoption projects.

Omar McIver is a Microsoft Certified Trainer and has more than 12 years of experience in developing enterprise-grade applications in regulated and global retail industries. He specializes in cloud-native development and application modernization. He is a certified Azure Solution Architect and FinOps Practitioner. His Udemy course on Practical OAuth, OpenID, and JWT in C# .NET Core has a rating of 4.5 stars. Omar continues to stay at the forefront of cloud-native development, with a keen focus on cost optimization, performance tuning, and highly scalable microservice architectures.

About the reviewers

Mikhail Filippov works at JetBrains as the team lead of Rider (cross-platform IDE for .NET). During 20 years of software development experience, Mikhail got deep technical expertise in various areas, including .NET runtime internals, ASP.NET MVC, MSBuild, and best practices for designing large-scale development infrastructures. Mikhail is a speaker at multiple events for developers and an active open source contributor. In his free time, Mikhail likes reading science fiction books and working on hobby projects using C# and Kotlin.

Jamie Taylor is a Microsoft MVP, software engineer, mentor, and podcaster with over a decade and a half of experience working with the .NET technology stack.

His primary focus is on his family, who inspire him every day. With a background in teaching, he focuses his energies on mentoring, writing incredibly high-quality code, and sharing his knowledge with everyone who will listen.

Outside of development and mentoring, Jamie is a keen podcaster - hosting three incredibly successful podcasts: The .NET Core Podcast, Tabs and Spaces, and The Waffling Taylors - and has collaborated with musicians, creating music and playing the bass guitar on a few recorded pieces.

Table of Contents

Part 1: Event-Driven Architecture and .NET 7

1

2

Part 2: Testing and Deploying Microservices

5

6

7

8

CI/CD Pipelines and Integrated Testing   147

9

Fault Injection and Chaos Testing   159

Part 3: Testing and Deploying Microservices

10

11

12

Service and Application Resiliency   217

13

Telemetry Capture and Integration   239

14

Observability Revisited   255

Assessments   283

Index   291

Other Books You May Enjoy   302

Preface

This book will guide you through various hands-on practical examples for implementing event-driven microservices architecture using C# 11 and .NET 7. It has been divided into three distinct sections, each focusing on different aspects of this implementation.

The first section will cover the new features of .NET 7 that will make developing applications using EDA patterns easier, the sample application that will be used throughout the book, and how the core tenets of **domain-driven design** (**DDD**) are implemented in .NET 7.

The second section will review the various components of a local environment setup, containerization of code, testing, deployment, and observability of microservices using an EDA approach.

The third section will guide you through the need for scalability and service resilience within the application, along with implementation details related to elastic and autoscale components. You'll also cover how proper telemetry helps to automatically drive scaling events. In addition, the topic of observability is revisited using examples of service discovery and microservice inventories.

You'll also learn the tenets of event-driven architecture, coupled with reliable design patterns to enhance your knowledge of distributed systems and build a foundation for professional growth. You'll also understand how to translate business goals and drivers into a domain model that can be used to develop an app that enables those goals and drivers, and identify areas to enhance development and ensure operational support through the architectural design process.

By the end of this book, you'll be able to identify and catalog domains, events, and bounded contexts to be used for the design and development of a resilient microservices architecture.

Who this book is for

This book will help .NET Developers and architects looking to leverage or pivot to microservices while utilizing a domain-driven event model.

What this book covers

Chapter 1, *The Sample Application*, explores the sample application, starting with an architectural overview and going into implementation details of entities, value objects, services, factories, and aggregates.

Chapter 2, *The Producer-Consumer Pattern*, reviews the fundamentals of the producer-consumer pattern (also known as the publisher-subscriber pattern) and how this pattern is implemented in the application through code examples.

Chapter 3, Message Brokers, covers details and implementation methods related to the use of message brokers within the larger ecosystem of the application. The term *"message broker"* will be defined, examples of technologies that act as message brokers will be discussed, and evaluations will be made to facilitate the choice of the appropriate technology to address the needs of the application.

Chapter 4, Domain Model and Asynchronous Design, covers the details of the domain model, including commands and events. It also covers the implementation of asynchronous events and when to use them within the application.

Chapter 5, Containerization and Local Environment Setup, examines how containerization is used to isolate functionality for build and deployment, and how to set up your local environment to enable this workflow.

Chapter 6, Localized Testing and Debugging of Microservices, takes your local environment further, building upon the environment setup and addition of the Dockerfile. You will learn how to test and debug your microservice in your local environment.

Chapter 7, Microservice Observability, reviews different types of observability at the service level, frameworks and technologies that enable observability, and how greater observability can be helpful in the application's ecosystem for both developers and operations.

Chapter 8, CI/CD Pipelines and Integrated Testing, reviews common **continuous integration** (**CI**) and **continuous delivery** (**CD**) strategies, practical implementations using GitHub Actions, and how to integrate baseline and regression testing into your pipelines.

Chapter 9, Fault Injection and Chaos Testing, reviews the concepts of fault tolerance and fault injection as it relates to software testing. More complex testing methods, such as stress and chaos testing, will be examined and you will learn how these techniques can be implemented to validate resilience and fault tolerance.

Chapter 10, Modern Design Patterns for Scalability, reviews different design patterns that aim to enable service scalability, as well as cover implementing scalability constructs in two popular deployment targets (Kubernetes and Azure API applications).

Chapter 11, Minimizing Data Loss, teaches specific techniques for minimizing or eliminating data loss as a result of scaling operations or as a function of increased service resiliency. Paradigms such as eventual consistency and guaranteed delivery will be reviewed along with a refresher on identifying data that may be susceptible to loss and defining how much if any, loss is acceptable.

Chapter 12, Service and Application Resiliency, helps you implement patterns, from what you've learned about scalability and data loss, that will guarantee service uptime, business continuity, and end-user satisfaction. Part of ensuring a consistent experience is generating meaningful messages to the end user which can be handled in a variety of methods.

Chapter 13, Telemetry Capture and Integration, examines different options for capturing application-level and service-level telemetry, and how to ensure that relevant information is captured without producing unnecessary noise or overhead. You will also learn how to pinpoint meaningful telemetry and aggregate that as opposed to aggregating everything, which can lead to confusion, large storage footprints, and a distrust of the information captured.

Chapter 14, Observability Revisited, in this chapter, you will learn about methodologies for publishing service metadata to your organization, cataloging and versioning microservice metadata, and how to promote the discovery of shared services within an organization.

To get the most out of this book

Software/hardware covered in the book	Operating system requirements
Visual Studio / Visual Studio Code	Windows, macOS, or Linux
.NET 7	Windows, macOS, or Linux
Azure Cloud Services	Windows, macOS, or Linux
Terraform	Windows, macOS, or Linux
Kubernetes (Cloud-based, Docker Desktop, MiniKube, k3s, or MicroK8S)	Windows, macOS, or Linux
Docker	Windows, macOS, or Linux
Docker Compose	Windows, macOS, or Linux

If you are using the digital version of this book, we advise you to type the code yourself or access the code from the book's GitHub repository (a link is available in the next section). Doing so will help you avoid any potential errors related to the copying and pasting of code.

Download the example code files

You can download the example code files for this book from GitHub at `https://github.com/PacktPublishing/Implementing-Event-Driven-Microservices-Architecture-in-.NET-7`. If there's an update to the code, it will be updated in the GitHub repository.

We also have other code bundles from our rich catalog of books and videos available at `https://github.com/PacktPublishing/`. Check them out!

Download the color images

We also provide a PDF file that has color images of the screenshots and diagrams used in this book. You can download it here: `https://packt.link/C9mhU`.

Conventions used

There are a number of text conventions used throughout this book.

`Code in text`: Indicates code words in text, database table names, folder names, filenames, file extensions, pathnames, dummy URLs, user input, and Twitter handles. Here is an example: "The Zookeeper configuration is not shown as it is identical to the first `docker compose` file."

A block of code is set as follows:

```
using OpenTelemetry.Metrics;
using OpenTelemetry;
using System.Diagnostics.Metrics;

var meter = new Meter("NameOfMeter");
var counter = meter.CreateCounter<int>(Name: "requests-
    received", Unit: "requests", Description: "Number of
      requests the service receives");
```

Any command-line input or output is written as follows:

```
dotnet add package System.Diagnostics.DiagnosticSource
dotnet add package OpenTelemetry.Instrumentation
  .EventCounters --prerelease
```

Bold: Indicates a new term, an important word, or words that you see onscreen. For instance, words in menus or dialog boxes appear in **bold**. Here is an example: "Be sure to uncheck the **Use controllers** checkbox to use minimal APIs if using the new project dialog from Visual Studio."

> Tips or important notes
> Appear like this.

Get in touch

Feedback from our readers is always welcome.

General feedback: If you have questions about any aspect of this book, email us at `customercare@packtpub.com` and mention the book title in the subject of your message.

Errata: Although we have taken every care to ensure the accuracy of our content, mistakes do happen. If you have found a mistake in this book, we would be grateful if you would report this to us. Please visit `www.packtpub.com/support/errata` and fill in the form.

Piracy: If you come across any illegal copies of our works in any form on the internet, we would be grateful if you would provide us with the location address or website name. Please contact us at copyright@packt.com with a link to the material.

If you are interested in becoming an author: If there is a topic that you have expertise in and you are interested in either writing or contributing to a book, please visit authors.packtpub.com.

Share Your Thoughts

Once you've read *Implementing Event-driven Microservices Architecture in .NET 7*, we'd love to hear your thoughts! Scan the QR code below to go straight to the Amazon review page for this book and share your feedback.

https://packt.link/r/1-803-23278-1

Your review is important to us and the tech community and will help us make sure we're delivering excellent quality content.

Download a free PDF copy of this book

Thanks for purchasing this book!

Do you like to read on the go but are unable to carry your print books everywhere? Is your eBook purchase not compatible with the device of your choice?

Don't worry, now with every Packt book you get a DRM-free PDF version of that book at no cost.

Read anywhere, any place, on any device. Search, copy, and paste code from your favorite technical books directly into your application.

The perks don't stop there, you can get exclusive access to discounts, newsletters, and great free content in your inbox daily

Follow these simple steps to get the benefits:

1. Scan the QR code or visit the link below

https://packt.link/free-ebook/9781803232782

2. Submit your proof of purchase
3. That's it! We'll send your free PDF and other benefits to your email directly

Part 1:
Event-Driven Architecture
and .NET 7

This part will cover the new features of .NET 7 that will make developing applications using EDA patterns easier, as well as the sample application that will be used throughout the book and how the core tenets of **Domain-Driven Design (DDD)** are implemented in .NET 7.

This part has the following chapters:

- *Chapter 1, The Sample Application*
- *Chapter 2, The Producer-Consumer Pattern*
- *Chapter 3, Message Brokers*
- *Chapter 4, Domain Model and Asynchronous Design*

1

The Sample Application

Over the past several years, the emergence of high-volume, scalable, event-driven applications has caused an interesting shift in application development. Complimentary design patterns have made writing and implementing event-driven architectures more appealing and have helped to reduce the learning curve when it comes to fully leveraging the elasticity and resiliency of cloud platform components. We will be taking a look at an application that utilizes event-driven architectures, implemented using .NET 7 and leveraging cloud-native applications and data constructs.

The purpose of this chapter is to outline the sample application we will be using throughout this book, along with the business drivers and goals it intends to satisfy. This will provide you with the opportunity to get a baseline understanding of the application's structure, source code, mechanics, and domains.

In this chapter, we'll cover the following main topics:

- Exploring business drivers and the application
- Architectural structures and paradigms
- Implementation details

Technical requirements

There are several prerequisites you will need to have an understanding of or have installed on your machine to use the code base and follow along with the examples. These include the following:

- Git
- Visual Studio or Visual Studio Code
- Docker
- Kubernetes
- Service-oriented architectures
- **Domain-Driven Design** (DDD)

We will be using an application that has been custom-developed and is included with the source code for this book. The primary platform we will be using to develop in will be .NET 7. All examples will use Visual Studio 2022 as the primary **integrated developer environment** (**IDE**). Either Visual Studio 2022 or Visual Studio Code will be required to develop .NET 7 solutions.

> **Important note**
> The links to all the white papers and other sources mentioned in this chapter are provided in the *Further reading* section toward the end of the chapter.

Exploring business drivers and the application

It's always a good idea to have a solid understanding of why an application exists, how it came to be, and what problems or opportunities it looks to solve. This application is a concept application that involves **Internet of Things** (**IoT**) devices, distributed event ingestion at scale, and facial recognition features. The primary market for this application is for turnstiles used at mass transit locations:

Figure 1.1 – Turnstiles in use at a transit station

In this scenario, the baseline events capture a simple count of customers who pass through the turnstiles at both the entrance and exit points of the mass transit system. Some drivers that contribute to the concept of the application, along with its need, include the following:

- To increase the visibility of equipment health and the need for proactive maintenance

- To allow integration for facial recognition sensors that can scan law enforcement databases for potential fugitives or persons of interest

- To manage costs associated with turnstile equipment, with options for expanding to fare payment interfaces

- To analyze transit usage, turnstile placement, and the need for additional units in high-volume areas

Having the ability to capture foot traffic related to the entrance and exit points of a transit station has several benefits. First, it can be used to understand how busy any one station is. Second, with extended use, the equipment can wear down and eventually break. With a line of sight into how many people are using the equipment, technicians can make educated decisions regarding when units might need to be serviced or ultimately replaced. This could also lead to the deeper monitoring of other components besides the turnstile unit, such as the payment interfaces. Some units might only have a ticket scanner, while others might have a ticket scanner and an electronic payment interface, where contactless payments using mobile devices can be used. The monitoring of normal usage, malfunctions, and scheduling the proactive servicing of those components could also be beneficial.

An additional use case could be that of transit scheduling and vehicle availability. Generally, the number of vehicles (such as trains, trams, buses, and more) any transit authority might have in its fleet is a direct result of them already monitoring customer traffic demands. Using data that has been captured in real time can help accelerate the analysis of needed schedule adjustments, fleet adjustments, or reductions in services for less-traveled stations.

The addition of facial recognition software to the equipment is not a hard requirement but does offer a value-add in the ability to potentially identify criminals at large or suspects who are wanted for questioning. With any artificial intelligence, it is essential to both program and operate with ethics and security in mind. While closed-circuit cameras and more advanced video surveillance equipment can be found in many transit stations, those cameras do not immediately notify anyone if a person has been recognized based on an alert or a bulletin issued by a law enforcement agency. Data collected during facial scans must be treated as personally identifiable information and must be purged if no match has been found.

Unpacking this a bit more, other potential drivers could come into play. For example, examining the business requirements for the application would add clarity. Looking at the domain model and any **domain-specific language** (**DSL**) associated with the requirements would help remove any ambiguity around what is meant by a customer, an order, an item, or even a payment method. Let's take a look at the domain model to get a better understanding of the layout of the different services, contexts, and aggregates.

Reviewing the domain model

The application's domain model describes the functional areas (domains) that live within the confines of the application. Each is developed using a ubiquitous language that everyone—from business analysts to senior leadership, to junior developers—can easily understand and relate to. *Figure 1.2* represents a simple domain model diagram that aligns to the structure of the application:

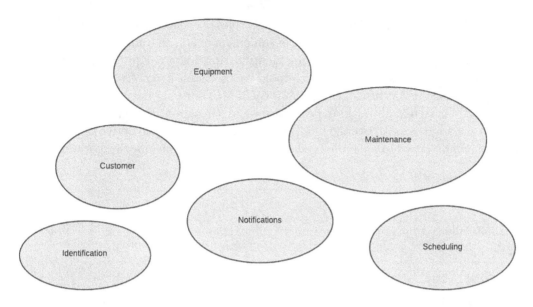

Figure 1.2 – A high-level domain model

The primary domains we will reference for this application are related to the primary pieces of functionality the application looks to offer. The following table offers a description of each domain:

Domain	Event and data scope
Equipment	Physical equipment implemented at a transit station
Identification	Data used to invoke facial recognition services
Maintenance	Maintenance requests (proactive or reactive)
Scheduling	Scheduling maintenance events; can also be extended to the service scheduling of transit vehicle services
Passenger	Passenger information, specifically for fare payment or optional membership services
Notification	Notifications that might be sent upon events being handled

Table 1.1 – Application functions

With these baseline domains defined, some simple rules of engagement can be derived. For example, a passenger could use a piece of equipment to enter a transit station while being run through facial recognition by the **Identification** domain. Equipment could raise an error noting a malfunction, which could then schedule a maintenance event. Equipment events such as turnstile operations could fire events per turn, allowing the aggregation of passenger throughput per turnstile and per station. These interactions can then be broken into areas of overlapping concern and, ultimately, help derive aggregate roots that are important to the model and the application. They include the following:

- Passenger
- Station
- Turnstile
- Camera
- NotificationConfiguration
- TurnstileMaintenanceSchedule
- CameraMaintenanceSchedule

Each of the aggregates will contain common properties such as the name and the ID. Some differences between entities and value objects related to the aggregates will be required, as each one will have its own requirements for data, as prescribed by the domain. *Figure 1.3* represents a high-level diagram of each aggregate, including properties (the list items), entities (the white rounded rectangles), and value objects (the green rounded rectangles):

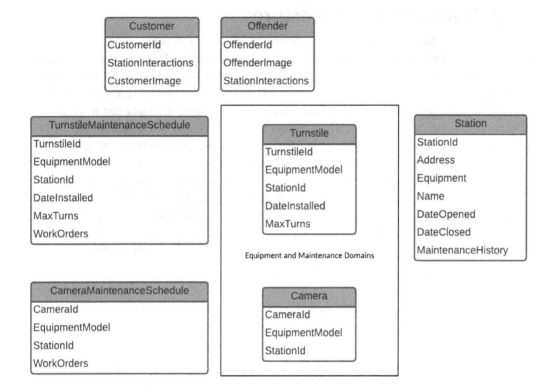

Figure 1.3 – A high-level aggregate view

Chapter 4, Domain Model and Asynchronous Design, dives deeper into the domain model, including a review of events and event handlers and asynchronous design.

With an understanding of the business relevance and the domain model that supports the business case, next, we can go one level deeper and examine some of the architectural structures and paradigms that help to define the event-driven nature of this application.

Assessing architectural structures and paradigms

Establishing an architectural baseline helps to drive decisions regarding how the application and its components will ultimately be implemented. Additionally, it also provides an opportunity to evaluate different patterns and practices with the ultimate goal of selecting a path forward. This section covers the overall architectural design of the sample application and some core tenets that enable the creation and consumption of events.

A high-level logical architecture

The solution is predicated on the use of hardware interfaces (such as equipment) that can communicate to hosted services in the cloud via a standard network connection. There is a hardware gateway (such as Raspberry Pi) that hosts simple write-only services, which will integrate using relevant domain services to record turnstile usage, facial recognition hits, and possible malfunctions with the turnstile or camera. Any user interface can interact with a common API gateway layer, which allows for data exchange without needing to know all the particulars of the available APIs. The backend runtime is managed by Kubernetes (in this particular case, AKS), with containers for each of the available domain microservices. Each of these microservices interacts with the event bus to send events. Then, the events are handled according to the domain's applicable event handlers. A reporting layer is used to access information captured via the event stream. SQL databases will be used to maintain the append-only activity log of events that come in via Kafka, and read models will be consumed from domain databases using read-oriented services.

The following reference diagram shows the logical construction of the application:

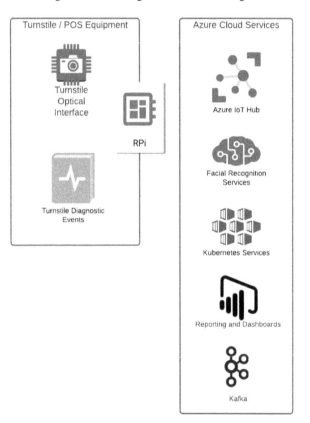
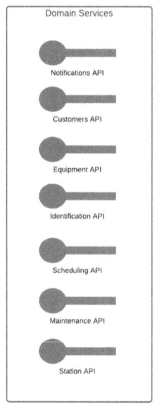

Figure 1.4 – A logical high-level reference architecture

The application uses the *Producer-Consumer* pattern to produce events, which are later consumed by components who need to know about them. You might also see this pattern referred to as *Publish-Subscribe* or *pub-sub*. The key point to take away from the use of this pattern is that any number of components could produce events containing relevant domain information, and any number of possible components could consume those events and act accordingly. We will dive into the producer-consumer pattern in much more detail in *Chapter 2, The Producer-Consumer Pattern*.

Digging down a layer, there are two technology architecture specifications that we will be using. One is for the device board inside the turnstile unit, which hosts the *Equipment* domain service. The other is the layout of the cloud components, as mentioned in the reference architecture in *Figure 1.4*. The high-level flow between the turnstile device and the cloud components is as follows:

- On the turnstile, after completing one turn, a message is sent to the equipment service indicating a completed rotation.

- The equipment service will send an event to the IoT hub with the results of the turnstile action.

- Using Kafka Connect, the message will be forwarded to Kafka, implemented within the Kubernetes cluster using the confluent platform.

- The event will be written to the appropriate stream.

- Any relevant event handlers will process the event.

  A more detailed diagram of the technology architecture can be seen in *Figure 1.5*, where both the turnstile unit and the cloud components are represented:

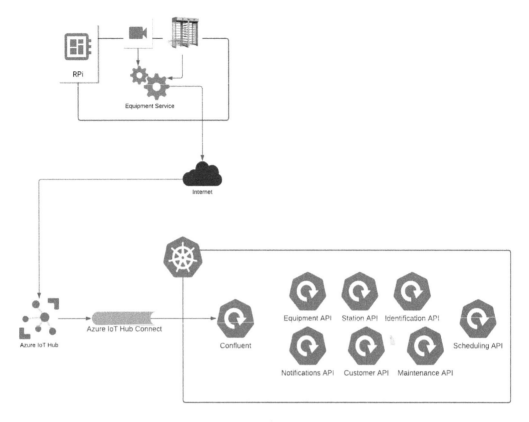

Figure 1.5 – The technology architecture for turnstile-to-cloud communication

Next, we will move on to the design of the event sourcing technique.

Event sourcing

Event sourcing is a technique that allows an application to append data to a log or stream in order to capture a definitive list of changes related to an object. One of the benefits of using event sourcing versus traditional **create, retrieve, update, and delete** (**CRUD**) methods with relational databases is that the performance can be tuned and increased at the service level, as the overhead of using CRUD methods is not needed. Also, it facilitates implementing a separation of concerns and the single responsibility principle, as outlined by the SOLID development practices (`https://en.wikipedia.org/wiki/SOLID`).

Another benefit of using event sourcing is its ability to achieve high message throughput while maintaining a high degree of resiliency. Technologies such as Kafka inherently allow for multiple message brokers and multiple partitions within topics. This design ensures that, at the very least, one broker is available to communicate with, and multiple partitions within a topic allow for data redundancy and scalability since Kafka will replicate partition data to each broker in the cluster. This enables multiple consumers to access or write data in parallel.

When using event stores with streaming capabilities, it enables you to debug point-in-time data and replay events to aid in debugging. For example, if an event has data that causes an error in the service code, you are fully able to go back to the point in time before that error was thrown and replay events to help identify potential bugs. Additionally, it can be used to perform "what if" testing. In some cases, normal use cases might have related edge cases that could either cause issues or introduce complexities that they were not originally designed for. Using "what if" testing allows you to go to a certain point in time and begin issuing new events that would correlate to the edge case while also monitoring application performance and potential failures.

Command-Query Responsibility Segregation

Command-Query Responsibility Segregation (CQRS) is a design pattern introduced by Greg Young that is used to describe the logical and physical separation of concerns for reading and writing data. Normally, you will see specific functionality implemented to only allow writing to an event store (commands) or only allow reading from an event store (queries). This allows for the independent scaling of read and write operations depending on the needs of the application or the needs of a presentation layer, either in the form of business intelligence software, such as PowerBI, or web applications accessible from desktop and mobile clients.

Details around how CQRS impacts the design of the application's domain services are covered in the next section. It's important to note that having that distinct separation of concerns is vital to leverage the pattern effectively.

Reviewing the implementation details

After looking at the patterns that will support the business use cases for the application, now, we can move on to the more specific implementation details. While some of the implementation constructs used in this solution will seem familiar, there are some technical details that might be new to you. We will be exploring several topics in this section, which are intended to prepare you for the journey ahead.

The Visual Studio solution topology

The solutions within the source folder are broken up by domain, with a separate solution for each. Additionally, there is a solution for core platform needs, such as marker interfaces to identify value objects, entities, aggregates, and other objects. The intent is to allow for each of the services to be run as an independent solution, which are eventually moved into their own repositories if so desired.

Each of the domains will have API services that can be communicated with. These projects in Visual Studio are not overly complex or even far from the general project template that is created when you create a new .NET Core API app. There are separate project types for queries, which read data, and commands, which affect data. Each domain will have a domain library, an infrastructure library, and test projects where applicable. Also, each domain will have a persistent consumer, in the form of an executable, that will run to enable listening for domain messages and handle those messages accordingly.

Solution folders will also be present to house Docker files, Docker Compose files, and any relevant **Infrastructure-as-Code** (**IaC**) or **Configuration-as-Code** (**CaC**) required to deploy the necessary components. Eventually, this will also be the location of the YAML file that defines the build and release pipeline.

> **Important note**
> The namespaces in each solution all start with a common acronym: **MTAEDA**. This stands for **Mass Transit Authority Event-Driven Application**.

Identity and Access Management considerations

Managing access to an application can be a daunting task. Many different options are available, from standalone implementations to platform-native solutions such as Azure Active Directory. Sometimes, the choice to go with an identity provider can be left with the application team; other times, it is driven by an enterprise strategy for authentication and authorization.

In this case, authentication will be handled at two layers. One layer is for transmitting events to applicable services, and the other layer is for users to log in and access management tools, such as dashboards and reports. As the dashboards and reports will be hosted in PowerBI, Azure Active Directory will be used to manage the authentication and authorization of those assets. For communication to the gateway and subsequent domain services for read and write operations, certificates will be used to govern traffic from the equipment to the gateway.

Event structure and schema

To help simplify and streamline event constructs, we have selected the CloudEvents open specification as the baseline for all events being transmitted. This allows you to capture relevant metadata about the operation while still sending over the event data itself. Additionally, using the CloudEvents schema enables you to potentially leverage platform tooling such as Azure Log Analytics and Azure Monitor. Of course, if your cloud target is different, there might be other ways the event schema could be useful. However, in this book, we will focus on the Azure cloud platform.

The schema for a CloudEvents schema is rather simple. There are fields for **Data**, **Subject**, **Type**, **Source**, **Time**, and **DataContentType**. They do not all require values; however, we will be using them to help better define the intent and content of each event we raise. It is entirely possible to not use this construct and still use the domains and domain services. The primary reason this design decision was made was to ensure there is consistency in the message format, along with a capacity to understand metadata associated with the event itself. *Table 1.1* illustrates the CloudEvent fields and how they will be used to contain pertinent information when an event is raised:

The CloudEvent property	Description	Mapping
Data	The content of the event being raised	Domain event; might contain information relevant to an aggregate or entity
DataContentType	The type of data being stored in the Data field	
Source	The source of the event	The domain method being invoked to raise the event
Subject	A friendly description of the intent of the event	Phrasing relative to the event's reason for being raised (Turnstile is locked)
Time	The time the event was raised	A timestamp correlating to when the event was raised
Type	The type of event being raised	`DomainEvent,` `DomainEventHandler,` `DomainCommand,` `DomainException`

Table 1.2 – The CloudEvent schema and field mappings

Local development and debugging

For local development, using Visual Studio is the easiest option to ensure any prerequisites for the solution can be installed and managed. Additionally, you can use Visual Studio Code, or even use GitHub CodeSpaces, to leverage a fully encapsulated development environment in the cloud.

If you are using Windows as your primary operating system, you will likely also leverage the **Windows Subsystem for Linux** (**WSL**), which allows for Linux-native builds and tooling to be directly run from Windows. In the event that any SDKs are missing, Visual Studio will alert you to that, and allow you to install them by clicking on a link next to the message.

There are a couple of different options that you can use to debug the application locally:

- Start debugging directly from Visual Studio (*F5*).

- Run the application using `docker compose` and attach to the Docker processes via Visual Studio.

- Deploy the application to Kubernetes and attach it to the application using the Kubernetes extension in Visual Studio.

New .NET 7 features

With the rollout of .NET 7, many improvements have been made to the underlying functionality offered along with language-specific updates. In this application, we will be taking advantage of some of the latest updates from a framework and language perspective. Language-wise, the implementations of minimal APIs and the asynchronous streaming of JSON data will come in handy for simplifying service implementations, and the ability to leverage Hot Reload will allow for faster and more meaningful debugging during the development life cycle.

Minimal APIs

One of the more exciting features in .NET 7 is a feature called minimal APIs. This allows you to develop an ASP.NET Core Web API app with very little code. The .NET team has worked on making the `using` statement a global construct—meaning that top-level statements, such as `using System` or `using Microsoft.AspNet.MVC`, are assumed to be required by all files within a Web API project and are not required in each file as a result. Additionally, the `Startup.cs` file is no longer required, as you can configure the app directly from the main `Program.cs` file. The following example code illustrates a code block that is valid and will create an ASP.NET Core Web API app when it is compiled:

```
var app = WebApplication.Create(args);
app.MapGet("/api/testing",(Func<IActionResult>)(() => { return
    new ContentResult() { Content = "Testing" }; }));
app.Run();
```

For a very simple API, you can map Get, Post, Put, Patch, and Delete operations directly in the Program.cs file, and they will be added to the routes for the Web API app. Additionally, you can call app.MapControllers() if you wish to keep controller code in separate files, as found in traditional Web API project layouts. On startup, the application will look for items derived from the Controller base class. If you choose this option, you will need to invoke the WebApplication.BuildConfig() method and pass in the build configurations, telling the application to add controllers to the configuration services, as demonstrated in the following code block:

```
var builder = WebApplication.CreateBuilder(args);
builder.Services.AddControllers();
var app = builder.Build();
app.MapControllers();
app.Run();
```

JSON transcoding for gRPC

While support for gRPC services was originally added in .NET 6, further improvements have been introduced to enhance the experience. Previously, in order to connect to a gRPC service for testing purposes, you had to build a client for that service and interact with it via the client. With the addition of JSON transcoding support, you can now launch a Swagger page that contains all of the available methods you are exposing via ProtoBuf, and perform tests against them. This doesn't replace the need to have a client built for communication purposes when deployed, but it does help the experience of testing locally.

Observability

With .NET 7, the integration with OpenTelemetry allows developers to leverage out-of-the-box instrumentation as well as telemetry exporters for popular site reliability platforms such as Prometheus and Jaeger. OpenTelemetry is a platform-agnostic framework that enables developers to expose both stack metrics (such as ASP.NET Core instrumentation) as well as custom metrics based on counters, histograms, and meters. While there is active work being done on these libraries, there are versions available that can be installed via NuGet and makes adding baseline telemetry capture straightforward.

Hot reload

One bit of functionality that has been present in other web development stacks for years but not in Visual Studio itself is the option to hot reload when debugging. For example, if you were to change a line of code in a controller, you would need to stop debugging, change the line of code, then resume debugging. With Hot Reload support in .NET 7, this is no longer an obstacle. In Visual Studio 2022, there is now a new icon that invokes hot reload once a change has been detected in the underlying source code.

Summary

This chapter provided an overview of the sample transit application, including the underlying business drivers, architectures, and implementation patterns. We have taken a quick lap around the domain model along with aggregates, entities, and value objects. Additionally, we have covered some key areas within the application's architecture, along with some specific implementation details, including new features in .NET 7 that will make development and debugging easier for us. All of these core topics will be covered in more detail in the coming chapters.

The next chapter takes a look into the producer-consumer pattern, which is an essential underpinning of the application and what helps event-driven systems work at scale. We will be looking at the underlying usage of this design pattern, how it benefits applications that operate at scale, how it is implemented, and how to validate that communications are properly being routed and sent.

Questions

Answer the following questions to test your knowledge of this chapter:

1. What potential insights can be gained when examining the business perspective behind an application?

2. Are there other domains that you can identify for the application that are not already listed in the primary domain model?

3. Are any of the aggregates misrepresented? Or do they contain information that might be irrelevant within the scope of the domain?

4. How is event sourcing different from using a relational database or NoSQL database to store and retrieve application data?

5. Is there an advantage to separating read operations from write operations?

6. What benefits can be gained by separating domain solutions from the overall application solution? Are there potential drawbacks to separating the domain solutions?

7. What other authentication and authorization mechanisms are available to secure access to reporting data and/or the write services that send data to Kafka?

8. Is using a standard schema for events, such as CloudEvents, unnecessarily complicating the overall design of the application? Why or why not?

9. What are some alternative implementations for these services aside from Docker or Kubernetes?

Further reading

- *Domain-Driven Design: Tackling Complexity in the Heart of Software* by Eric Evans: This is available at `https://www.domainlanguage.com/`.

- *Getting Started with EventStore*: This is available at `https://developers.eventstore.com/server/v21.6/docs/introduction/#getting-started`.

- *Implementing Domain-Driven Design* by Vaughn Vernon: This is available at `https://vaughnvernon.com/`.

- *CQRS Documents* by Greg Young: This is available at `http://cqrs.files.wordpress.com/2010/11/cqrs_documents.pdf`.

- *CloudEvents: A specification for describing event data in a common way*: This is available at `https://cloudevents.io/`.

- *ASP.NET Core Updates in .NET 7 Preview 4* by Daniel Roth: This is available at `https://devblogs.microsoft.com/dotnet/asp-net-core-updates-in-dotnet-7-preview-4/`.

2

The Producer-Consumer Pattern

When deciding on a communication pattern for an application, it is normal to look at the circumstances under which components might communicate with one another. Typical web applications follow a request/response pattern, where a request is made to the server, and a response is returned to the client and handled appropriately. For applications looking to handle high throughput in those communication channels, a standard request/response pattern is not ideal.

With hundreds or thousands of requests being sent per second, the application would be slowed to a halt as every request is met with a corresponding response. Another example is that of a long-running process. There might be an operation that gets kicked off by one event but doesn't complete until several hours later. Blocking the thread running that process would limit execution ability, and ultimately, the system only needs to know when the process completes. Scenarios such as this call for a more resilient and asynchronous messaging pattern. This chapter intends to cover one such pattern, the **Producer-Consumer pattern**, while illustrating the following focus areas:

- Examining producers and consumers
- Exploring implementation details in code
- Reviewing implementation details in infrastructure

By the end of this chapter, you will have a clear understanding of the following:

- The premise behind the producer-consumer pattern, how its use enables **event-driven architecture** (**EDA**), and why the pattern is so widely used

- Reviewing and understanding the implementation of the producer-consumer pattern through code examples that tie back to the sample application

- Different infrastructure mechanisms that can be leveraged to easily enable producer-consumer patterns

Technical requirements

You will find all the code examples for this chapter available in the corresponding folder on GitHub at `https://github.com/PacktPublishing/Implementing-Event-Driven-Microservices-Architecture-in-.NET-7/tree/main/chapter02`.

> **Important note**
> The links to all of the white papers and other sources mentioned in this chapter are provided in the *Further reading* section toward the end of the chapter.

Examining producers and consumers

The producer-consumer pattern might be more familiar under a different moniker—*Publish-Subscribe* or *pub-sub*. While the terminology may differ, the intent behind the messaging pattern is the same. There are components that produce (or publish) events, and there are components that consume (or subscribe to) events. As mentioned in *Chapter 1*, *The Sample Application*, different mechanisms can support event-based messaging patterns. We have already established that we will be using Kafka as the primary mechanism for sourcing events, but if you've never used streaming technology before, it might be challenging to understand how streaming is relevant to communication patterns. Let's look at a couple of real-world examples, one of which happens to be the core use case for the sample application.

Relating to real-world examples

Developers understand the concept of communication patterns between components, as they enable those patterns while writing a particular application. An extremely popular pattern, which has been employed for decades, is that of request/response. Most people understand this pattern naturally, as it follows a similar pattern to how we, as humans, communicate. If you are speaking with someone and ask a question, you can generally expect a response from them. While that response might not be exactly what you expect or anticipate, you will still receive a response to your query.

For those who struggle a bit with understanding how producers and consumers work, a simple example can be seen in the form of an employee suggestion box. Employees write suggestions on a slip of paper and place it in a box that is generally locked. The suggestions go in but are not retrievable by anyone other than the person who holds the key to the box. These suggestions could be collected right away, at the end of the day, or at a specific time during the month. The important part of this interaction is that the person with the key to the box—typically, someone who is in management—collates those suggestions and brings them back to the group. Some might not be implemented, but the premise is that the suggestions will be read and potentially acted upon by those who can make changes in the workplace. The employees writing those suggestions are the producers, wherein, they produce suggestions to make the workplace more enjoyable or more efficient. In this case, the consumer would be the manager or supervisor, who collects those suggestions and acts on them, handling the suggestions as is appropriate for the workplace.

An additional point to note would be the consistency with which those suggestions are collected. A common theme in applications using the producer-consumer pattern is that of **eventual consistency**—meaning data will be sent and processed eventually, based on the configuration settings or requirements that specify how current data must be for a particular domain. If the suggestions are collected at the end of each day, this indicates a faster data consistency frequency than those suggestions that are collected once a month.

Next, we come to the sample application and its primary use case, which is monitoring the usage of turnstile equipment during daily operations at a mass transit station. The equipment raises events when it is used, when something goes wrong, or when the mechanism locks up. These events are sent to a **Kafka topic** (similar to a message queue) via a **message broker**, and consumers who are interested in those events will subscribe to that topic. Once an event hits the topic, the consumer brings it in, and an event handler will be run to process that event in accordance with domain rules. This may result in a write operation to a database, an entry in an application log file, or even in more events being raised. While more complicated than our first example, it does follow the same general behavior. The differentiator here is that producing, consuming, and handling events can occur at low or high numbers and these activities are governed by the business logic within the domain itself.

> **Important note**
> Message brokers will be covered in greater detail in *Chapter 4, Domain Model and Asynchronous Design*.

Each of these examples helps you to illustrate how interactions between producers and consumers can be found in everyday life. Next, we'll examine how this pattern enables event-driven architectures to be successful.

Enabling event-driven architectures

As a reminder, many event-driven systems are designed with the understanding that large volumes of messages will be produced and consumed at any given point. How the messages are produced and consumed needs to be highly **scalable**, **resilient**, and **fault-tolerant**. Otherwise, the structural integrity of the application is invalidated since there is no guarantee that any messages produced will be consumed by the components that are waiting for them.

Scalability can be addressed by a variety of different mechanisms. Certain services allow for an autoscaling feature to be leveraged, creating additional copies of the service to handle an increase in requests or usage. This is important because it ensures performance regardless of the usage pattern of the service—within reason. If the upper limit of the autoscale function is hit and the flow of messages is still relatively high, there could be performance implications if additional measures are not put into place. With the producer-consumer pattern, there is no explicit expectation of a response to any message produced. As mentioned earlier, the consistency of the system does not have to be immediate, which allows the consumers of those messages to process them as per their requirement, as long as they are processed.

System resiliency can be increased with the help of scalability, but other controls need to be in place to ensure the system is truly resilient. Software design patterns that are purpose-built to handle cloud architectures, such as those found on the **Azure Architecture Center** website, can lend additional safeguards to the system. For example, the circuit-breaker pattern can be implemented in code to safeguard against large surges of events coming in. When the threshold is hit, the breaker is "tripped" or left in an open state, allowing for a scaled throttling of subsequent events. Another pattern that is very commonly used is the **retry pattern**, where a specific operation can be retried a certain number of times, and at different intervals, if so desired. Often, technologies that support producer-consumer patterns have built-in retry logic. Kafka is one such technology.

Fault tolerance is the ability of a system to continue operating in the event of a critical fault or failure. As mentioned earlier, software patterns can also help the system to guard against faults. There will be certain faults that are transient and wide in scope that you might not immediately anticipate. When developing for the cloud, there is a notion that outages can, and will, occur. Regional services, entire data center regions, or global services can experience outages that last from seconds to days.

For every cloud service you plan to use, you should have a contingency plan for that cloud service not being available. This can lead to a deeper discussion around **business continuity** and **disaster recovery** scenarios, which are beyond the scope of this book. Conceptually, however, it's generally a good practice to have a plan in place to handle cloud service failures, whether through geographically distributed components or local operations on machines or equipment that can be persisted locally until the services come back online.

Often, accounting for these characteristics is a daunting task for application developers, as many are used to working in an environment where the underlying infrastructure is relatively static and stable. Along with the ramp-up time needed to account for these operational concerns, the adoption curve for truly leveraging the producer-consumer pattern can be steep for some.

Understanding the adoption curve

As we've seen in the previous subsections, having a solid understanding of the why behind using the producer-consumer pattern is just as important as the how, when, and where. From the developer experience perspective, moving toward this communication design pattern can bring a host of challenges. Not everyone will learn and adapt to all patterns at the same rate. The curve can be greater if the application already exists and refactoring or rearchitecting it to adopt this pattern is introduced into the development life cycle.

Tackling the learning curve starts with a conceptual understanding of the pattern itself. This can be conveyed through the examples listed earlier in this chapter, or through instructor-led sessions that are specifically tailored for application developers. There is plenty of additional online content via free sites, such as Microsoft Learn, which is intended to guide you through understanding concepts and design patterns that are meant for cloud-native development. Remember, this can be fundamentally different from the development practices you or your team are used to, and ensuring enough time is taken to absorb, experiment, and learn this content will help to solidify your ability to develop event-driven applications. Experimenting or hands-on learning is often beneficial. In the next section, we will talk through a high-level example of how to write a producer, a consumer, and an event handler, and then run those services to demonstrate how producer-consumer communication occurs.

Exploring implementation details in code

There are several ways in which the producer-consumer pattern can be implemented in code. Some samples can be downloaded and run, along with custom implementations that you might find in domain libraries or core (shared) libraries. First, we will look at sample code that uses the `Confluent` libraries for Kafka, which provide a simple and easy-to-understand abstraction layer for interacting with Kafka . The source projects for each of the components (producers and consumers) can be found in the chapter folder on GitHub, along with a bootstrap script to download and run Kafka locally. Please ensure that you have Docker installed, as the install is dependent upon running `docker compose`. The start order for this example code will be to run `docker compose up`, then to right-click on the `consumer` project and select **Debug** > **Start New Instance**, then to right-click on the `producer` project and select **Debug** > **Start New Instance**.

The producer code

We're going to start with the producer code, which can be set up using the **minimal API** feature in ASP.NET Core. To begin, create a new project in Visual Studio using the empty ASP.NET Core application template. *Figure 2.1* shows the template after utilizing a search with **C#** set as the language and **Web** set as the project type:

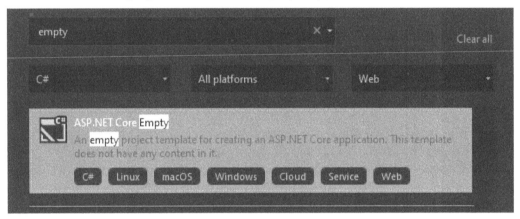

Figure 2.1 – The new project dialog in Visual Studio showing the empty ASP.NET Core project template

Next, we want to add some references to the project to enable the Swagger documentation of the service, the hosting of background services in the service, and the dependency injection mechanism in ASP.NET Core to resolve types appropriately. To do this, you can right-click on the project and select **Manage NuGet packages**. Alternatively, you can run the following code to install the packages within the project:

```
dotnet add package Confluent.Kafka
dotnet add package Microsoft.Extensions.Hosting
dotnet add package Microsoft.Extensions.DependencyInjection
dotnet add package Swashbuckle
```

```
dotnet add package Swagger
```

When you open the main `Program.cs` file, notice that there are no namespace declarations, no main method definitions, and no `using` statements. This is because the **top-level program** feature is employed to make the main code file concise and easy to read. The `Startup.cs` file is no longer required either, which makes setting up a basic web API project very simple and much quicker than previous framework versions. The following code is an example of what will be in your `Program.cs` file when you first create the project:

```
var builder = WebApplication.CreateBuilder(args);
var app = builder.Build();

if (app.Environment.IsDevelopment())
{
    app.UseDeveloperExceptionPage();
}

app.MapGet("/", () => "Hello World!");
app.Run();
```

We need a way to allow this service to produce a message that can be consumed by another program, which we will be building later in this chapter. This production component will need to be available to any mapped routes within `Program.cs`. One abstract class that is included in the `Microsoft.Extensions.Hosting` library is the `BackgroundService` class. This allows you to create a class that can interact with your Kafka instance behind the scenes, without impacting the primary thread of the web API process. During the API startup process, you can create a scoped instance of the background service that will be available to your controller methods via dependency injection.

To allow this binding to occur, we want to write an interface that implements the `IHostedService` interface. Using the following stub code, `IProducerService` would look like this:

```
interface IProducerService: IHostedService
{
    Task Send(string message);
    Task SetTopic(string topicName);
}
```

To hook in a concrete and scoped instance of this `IProducerService` interface, we will register it with a scoped instance as follows:

```
builder.Services.AddScoped<IProducerService,
  producerService>();
```

The producer component will need to leverage the `IProducer` interface from the `Confluent.Kafka` library alongside the `ProducerConfig` class, allowing for startup information to be passed to the `ProducerBuilder` class, which will generate a concrete implementation of `IProducer`. Following this, we will set up an instance of the `ProducerConfig` class in `Program.cs`, allowing us to read information from the `appSettings.json` file to populate that configuration by binding a section of the config to the variable. Then, we will register the instance of the `ProducerConfig` class as a singleton, allowing us to use it via dependency injection:

```
var producerConfig = new ProducerConfig();
...
var ProducerTopic = builder.Configuration.
  GetValue<string>("Topic");
builder.Configuration.Bind("ProducerConfig", producerConfig);
...
builder.Services.AddSingleton<ProducerConfig>(producerConfig);
builder.Services.AddSingleton(ProducerTopic);
```

Next, we will need to add a route to allow the POST action of a message to an endpoint. This method will contain some basic error handling and, for now, will deal only with a string-based message. This route will then call `ProducerService` and `await` the return of the `Send()` method. The following code can be put between the first `app.MapGet` function and the `app.Run()` function:

```
app.MapPost("/send", async (http) => {
    var message = "";
    using (StreamReader sr = new
      StreamReader(http.Request.Body))
    {
        message = await sr.ReadToEndAsync();
    }

    var svc = http.RequestServices
      .GetRequiredService<IProducerService>();
```

```
    await svc.SetTopic(ProducerTopic);
    try
    {
        if( message == null)
        {
            throw new InvalidDataException("Must have a
                message body");
        }
        await svc.Send(message?.ToString());
    }catch (Exception ex)
    {
        await http.Response.BodyWriter.WriteAsync(
            System.Text.Encoding.ASCII.GetBytes( ex.Message));
        http.Response.StatusCode = 500;
    }
    http.Response.StatusCode = 200;
});
```

We also need a place for the produced message to go. As mentioned previously, Kafka operates using the notion of topics. For this example, all we need to focus on is a topic where we can put the message being sent. You might have noticed that we have an additional configuration binding for the property of Topic. Entering a value for this in the appsettings.json file will allow you to set the name of the topic to which the producer service will write messages.

As a reminder, all of the sample code can be found in the chapter folder in the GitHub repository.

The consumer code

The consumer code is similar to the producer code in that it also employs the BackgroundService abstract class. The difference with the consumer service we will write is that it uses the CancellationToken value passed into the ExecuteAsync method to determine whether it should stop the service or not. If the token is not a cancellation request, the consumer service will continue to run in the background. Before we get into the main program loop, let's take a look at the consumerService class:

```
    internal class consumerService : BackgroundService,
        IConsumerService
```

```
    {
        private readonly IConsumer<int, string> consumer;
        private readonly string topicName;
        public consumerService(ConsumerConfig config,
          string topic)
        {
            consumer = new
              ConsumerBuilder<int,string>(config).Build();
            topicName = topic;
        }
        public async Task Receive()
        {
            await ExecuteAsync(CancellationToken.None);
        }
        protected override async Task
          ExecuteAsync(CancellationToken stoppingToken)
        {
            while (!stoppingToken.IsCancellationRequested)
            {
                consumer.Subscribe(topicName);
                var result =
                  consumer.Consume(stoppingToken);
                Console.WriteLine(result.Message.Value);
            }
        }
    }
```

As you can see, in this example, the ExecuteAsync method does the lifting. As long as the token is not indicating a request to cancel, the while loop will continue to run, allowing the consumer to subscribe to the topic name that you supply, read the message, and write the contents of the message out to the console.

Setting up the main program will be slightly different. Since this is not an ASP.NET Core Web API, there will need to be some extra considerations for loading the application configuration settings and initializing the consumerService class. Here, we can leverage items in the Microsoft.Extensions.Hosting library, specifically around how to configure and start the console application. In the sample code for the consumer project, the `Program.cs` file contains a standard `main` method, but it also contains the `CreateHostBuilder` method. The latter method allows you to set configuration sources and other bindings that we set up inline for the producer service code. The following is the `CreateHostBuilder` method for the consumer executable:

```
static IHostBuilder CreateHostBuilder(string[] args) =>
  Host.CreateDefaultBuilder(args).ConfigureAppConfiguration
  ((hostingContext, configuration) =>
            {
                configuration.Sources.Clear();
                IHostEnvironment env =
                  hostingContext.HostingEnvironment;
                configuration
                        .AddJsonFile("appsettings.json",
                            optional: true, reloadOnChange:
                            true)
                        .AddJsonFile($"appsettings
                        .{env.EnvironmentName}.json",
                            true, true);
                IConfigurationRoot config =
                  configuration.Build();
                _config = new ConsumerConfig() {
                  BootstrapServers =
                  config.GetValue<string>("KafkaServer"),
                  GroupId =
                  config.GetValue<string>("DefaultGroupId")
                  };
                topicName =
                  config.GetValue<string>("Topic");
        });
```

There are a few things happening here:

- There is a call to `CreateDefaultBuilder`, passing in the command-line arguments from the `Main` program method.
- Two placeholders are passed into the `ConfigureAppConfiguration` method for the hosting context and the configuration itself.

- JSON configuration files are specified as the source of the application's configuration settings. They are listed as optional based on the arguments passed into `AddJsonFile`; however, they will be required as that is where the topic name and the `ConsumerConfig` values will be pulled from.

- The private variables for the `ConsumerConfig` value, the default group ID, and the topic name are loaded from the JSON configuration files and set to static variables for later use. The group ID is significant for consumers of Kafka topics, as Kafka will keep track of what has (and has not) been read from a topic by a specific group. You can create multiple consumers that share the same group ID, and Kafka will automatically track what has been consumed by the group—not just the consumer application instance.

With that configuration taken care of, we can now focus on the `Main` method. It is fairly light in content as it only needs to initialize the application host, instantiate the `consumerService` class, and wait for the application to run:

```
using IHost host = CreateHostBuilder(args).Build();
var service = new consumerService(_config, topicName);
await service.Receive();
```

It's important to make sure the call to `Receive` is made before the call to `host.RunAsync()`, as the `Receive` method starts the background service. If the call is made after `RunAsync`, the consumer service might not be activated, and messages will not be consumed from the topic.

The intention is to run the producer code and the consumer code simultaneously, allowing you to send messages to the producer API and view the consumer retrieving the message and displaying its contents in the console. You can send messages to the API via `curl` (Linux) or `Invoke-RestMethod` (PowerShell or PowerShell Core). A sample PowerShell command could look like this:

```
Invoke-RestMethod -Method Post -Uri "https://localhost:5001/
   send" -UseBasicParsing -Body "This is a test"
```

However, we're not quite done with the consumer. Writing to the console directly from the `Receive` method is a quick way in which to observe results from the `consumerService` class, but we also want to ensure we're utilizing event handlers to process any messages or events received. Establishing this pattern now will pay dividends later on when the domain code is written.

Event handling

Generally, event handlers can be expected to be used in the consumer code. While the handler objects themselves are normally kept in the main domain library, the consumer implementation will pull in those event handlers, generally in a background process. Event handlers do not need to be terribly complex in nature, as we will see with the following code.

To get started, we will create a new class called `MessageReceivedEventHandler`, which will take the original console output call from the `Receive` method in the consumer code and abstract it into a separate event handler. This pattern will be employed in further detail, as we dive deeper into asynchronous programming, in *Chapter 4, Domain Model and Asynchronous Design*. For this example, the event handler class will have only one method, `Handle`, which will write the contents of the received message to the console:

```
internal class MessageReceivedEventHandler
{
    public async Task Handle(ConsumeResult<int,string>
        result) {
        await Task.Run(() => {
            Console.WriteLine(result.Message.Value); });
    }

}
```

Now, we need to remove the console output line from the `ExecuteAsync` method in the `consumerService` class and replace it with a new `await` statement:

```
while (!stoppingToken.IsCancellationRequested)
{
    consumer.Subscribe(topicName);
    var result =
        consumer.Consume(stoppingToken);
    await new MessageReceivedEventHandler()
        .Handle(result);
}
```

To test this new addition, run the producer and consumer projects in debug mode, and issue a command in the terminal of your choice to send a new message to the producer API. You should see the output appear in the consumer's console window.

Having run through the code implementations of the producer and consumer clients, it's time to gain more understanding of the implementation details around the infrastructure we will need.

Reviewing implementation details in infrastructure

In the last section, we focused slightly on the infrastructure provisioning of Kafka, although there is far more to setting up Kafka than simply running a script. To understand the producer and consumer examples, we made some assumptions about how to interact with Kafka and, specifically, a topic.

Topics

Topics are not unique to Kafka but are a construct that producer-consumer patterns leverage to store and retrieve messages relevant to a specific domain or grouping of events within a domain. Normally, topics are scoped to a specific subset of data. For example, with the domain model of the MTAEDA application, you m expect to find a topic for equipment, stations, and scheduling, among others.

Events and messages (known as records in Kafka) are written in an append-only fashion. This means that each record is immutable upon being written to a topic. Any changes that are required have to be appended to the end of the topic. This allows for the sequential reading of records, if needed, by the consumers.

Topics in Kafka are also comprised of partitions. Much like other data persistence technologies, partitions can be used to store events or messages that all have a common key. This ensures all relevant information for a particular key will be found within a specific partition. Enabling multiple partitions allows for greater distribution of datasets and, potentially, faster search capabilities when using that common key as a search term.

Additionally, partitions can be used as a means of load balancing across multiple brokers, as partitions themselves are viewed as parallel units. It can also add complexity to the solution—in some cases, you might not want to use multiple partitions because you aren't using a specific key value. It is possible to write a message or event to a topic with `null` as the key value. If this approach is taken, another common identifier will need to be used to associate events with one another if searching directly in a topic (if that association is required). If no key value is used, each record will be written to the end of a partition using a round-robin assignment.

Creating topics in Kafka is straightforward. You can create a topic by doing the following:

- Using a configuration setting within your Kafka broker (`auto.create.topics.enable`).

- Running a script to create a topic; can be done by downloading the latest release of Kafka from the Apache website and using the supplied scripts.

The first option will automatically create a topic for you using the defaults set in the Kafka broker, which normally defaults to one partition. Using the second option, you can create topics with more fine-grained configuration. There are also options for setting the partition count, the retention time, the replication count, and more.

> **Important note**
>
> Here is a quick note about the retention of records in Kafka. By default, Kafka will retain records for seven days. This can be adjusted based on application or storage considerations. As with any messaging platform, it should not be assumed that records will be retained indefinitely within Kafka. This is generally why many event-driven applications will use an alternative data store to capture history for each domain, and event handlers will be leveraged to write those records to the appropriate storage medium.

Streams and tables

Topics alone are not the only area in which records can be read or written. Kafka also has the concept of **streams** built in. In essence, streams are exactly what they sound like—a stream of information. Kafka provides a stream API that allows you to transform records being written into input topics and place those transformed records into output topics. A stream can be programmatically created based on a specific topic.

On the other hand, **Tables** are constructs that use data made available by streams and present that data in a specific and intentional way. Using common techniques such as **mapreduce** alongside Kafka-centric operations, you can create transformations that enrich or compact records with the intention of using another stream to write those records to output topics.

While these constructs are important and valuable as utilities within Kafka, you might find their use within domains as an augmentation to event handlers. Kafka does offer a connect API that allows you to set up connections with a variety of different destination systems and data stores. If the plan is to only consider and use Kafka as a messaging platform, using streams and tables along with the connect API could make a lot of sense.

For our example application, while Kafka is what we are using, there are abstractions in place that allow us to use different platforms for messaging and persistence, leaving the logic to the event handlers within the domain.

Aggregate storage

In the domain context, we have already identified several different possible aggregates that would need to be tracked. A common method for tracking changes to an aggregate when using a platform such as Kafka would be to have a topic for each aggregate. Events related to each aggregate can be handled by specific event handlers. Those handlers can then update the data stores as needed. *Figure 2.2* illustrates an example of aggregate-based topics in Kafka:

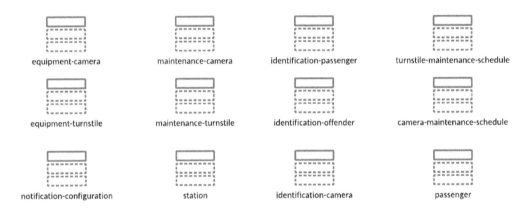

Figure 2.2 – An example topic layout by aggregate root

This is not to say that other events in non-aggregate topics could not impact data in those data stores; having a separate aggregate topic allows for a centralized way to process events related to multiple parts of an aggregate while still allowing for other event types to occur organically.

Let's take a look at an example using one of the domains in the example application. One of the aggregates within the equipment domain is the turnstile. The turnstile aggregate contains a few different pieces of information but does not contain items such as maintenance records, since those records are a part of the maintenance domain. However, it does contain a status field. In the event of the turnstile becoming locked, an event handler would need to update the backing data store with the new status and then produce an event requesting that a new maintenance record be created for the turnstile. Once the maintenance record has been created, the turnstile's status can be updated to reflect that the maintenance staff has been alerted. Once maintenance is complete, an update to the maintenance record could trigger a status update for the turnstile, indicating it is now functional. This example is illustrative of further integrations that will be explored when the domains are studied in more detail.

Summary

In this chapter, we covered a great deal of ground discussing the producer-consumer pattern. Not only did we define what a consumer is and what a producer is, but we also looked at how to implement messaging functionality in .NET 7, allowing both of them to interact with Kafka. We explored some features of the messaging platform infrastructure along with Kafka-specific features that allow for storing, transforming, and forwarding records.

With the knowledge of the producer-consumer pattern, as well as working on the code and infrastructure to support that pattern, you have established a good base understanding of how this messaging pattern can be implemented. Our next chapter centers on a specific construct within event-based messaging platforms—the message broker. The knowledge you've gained in this chapter will enable you to dive deeper into brokers and gain an understanding of how they are used in the larger scheme of event-driven systems.

Questions

Answer the following questions to test your knowledge of this chapter:

1. In the producer-consumer pattern, is both a producer and consumer required for the pattern to work?

2. Are there obstacles to learning or adopting the producer-consumer pattern for software engineers? If so, what are some examples?

3. What are the benefits of using minimal APIs in .NET 7? What are the drawbacks?

4. Which is the primary library used to facilitate the use of Kafka in the example code?

5. Which base class is used to construct the producer and consumer service classes?

6. Can configuration files such as `appsettings.json` be used to configure complex objects such as `ProducerConfig` and `ConsumerConfig`?

7. Is it generally a good idea to create topics with multiple partitions? Why or why not?

8. What's the difference between a stream and a table in Kafka?

Further reading

- *What Do You Mean by "Event-Driven"?* by Martin Fowler: This is available at `https://martinfowler.com/articles/201701-event-driven.html`.

- *Azure Architecture Center* by Microsoft: This is available at `https://docs.microsoft.com/en-us/azure/architecture/`.

- *Background tasks with hosted services in ASP.NET Core* by Jeow Li Huan: This is available at `https://docs.microsoft.com/en-us/aspnet/core/fundamentals/host/hosted-services?view=aspnetcore-5.0&tabs=visual-studio`.

- *Minimal APIs* by Microsoft: This is available at `https://minimal-apis.github.io/`

- *Apache Kafka Quickstart on Docker* by Confluent: This is available at `https://developer.confluent.io/quickstart/kafka-docker/`.

- *Official Apache Kafka Documentation* by the Apache Foundation: This is available at `https://kafka.apache.org/documentation`.

- *Kafka Producer/Consumer Tutorial* by Red Hat: This is available at `https://redhat-developer-demos.github.io/kafka-tutorial/kafka-tutorial/1.0.x/03-consumers-producers.html`.

- *Introduction to Kafka Topics and Partitions* by Coding Harbour: This is available at `https://codingharbour.com/apache-kafka/the-introduction-to-kafka-topics-and-partitions/`.

3
Message Brokers

As we saw in *Chapter 2, The Producer-Consumer Pattern*, the producer-consumer pattern is a highly effective way to handle messages at scale. Some of the scalability benefits can be seen in how programs can implement this pattern, while others can be seen in the infrastructure that enables the pattern. This chapter intends to dive deeper into message brokers while covering the following topics:

- What is a message broker?
- Inspecting messaging protocols, schemas, and delivery patterns
- Implementing message broker technologies

By the end of this chapter, you will be able to do the following:

- Understand the concept of a message broker and how it is used in the scope of the **event-driven architecture (EDA)**.
- Know how to use schemas for messages, various protocols that are used to send and receive messages, and their delivery options
- Know how to use specific implementations of message brokers and how to decide on the right technology

Technical requirements

You can find the code examples for this chapter in this book's GitHub repository at `https://github.com/PacktPublishing/Implementing-Event-Driven-Microservices-Architecture-in-.NET-7/tree/main/chapter03`.

> **Important note**
> The links to all the white papers and other sources mentioned in this chapter are provided in the *Further reading* section at the end of this chapter.

What is a message broker?

Simply put, a message broker is a piece of middleware that facilitates (or brokers) communication between two systems. While this sounds minimalistic, this basic concept can be expounded on for use cases such as the producer-consumer pattern, among others. Given the versatility of this type of middleware, it can be tempting to add additional processing features. Knowing when to add functionalities, such as data enrichment or transformation, is key to preventing product lock-in with a particular broker.

By means of **domain-driven design** (DDD) standards, keeping the implementation flexible while staying prescriptive on the interface functionality will yield the most consistent results. While you may not encounter a domain project that has 10 different concrete implementations of a broker or even a repository, it's not uncommon to have a few different options available. Often, you will see implementation classes for not only brokers but at least one type of persistence store; whether that's a cache, a database, or even a flat file.

Terminology aside, there are a vast array of message brokers out there. Some are free to use, while others are commercial software that requires licensing and payment. Regardless of pricing, though, brokers tend to fall into three main areas:

- Queue-based technology
- Cache-based technology
- Stream-based technology

Let's look at each type of broker, along with some general examples that fall into each category.

Queue-based technology

First, we'll examine brokers that are queue-based, meaning there is a conceptual construct that defines a queue or list where the messages sit. Processing those queued messages (also known as **dequeuing**) pulls them from the current queue, performs some sort of action on them, and yields a result that can be placed in another queue for continued processing.

The components that perform this processing could be executables, scripts, web services, microservices, or even built-in functionality with the broker itself. While discussing product-specific capabilities is beyond the scope of this book, the following products and their related documentation pages discuss what's possible with queue-based message brokers:

- **RabbitMQ**: An open source message queue and broker that allows you to use regular queue-based messaging and **advanced message queueing protocol** (**AMQP**) message routing.

- **ActiveMQ**: A Java-based platform that uses AMQP message routing.

- **Azure Service Bus**: A cloud-native message queue and relay specifically implemented in Microsoft Azure.

- **WebSphere Message Broker**: A message broker and transformation engine authored by IBM. This is commonly used with IBM MQ.

- **IBM MQ**: A messaging platform that heavily utilizes queues and topics for message storage and routing, authored by IBM.

While queue-based technologies may be more immediately familiar to some, other storage options are available. Next, we will discuss the concept of caching and some related technologies that support it.

Cache-based technology

Cache-based brokers tend to rely on underlying technologies that are used to cache data. Using caching mechanisms for critical messages may not be the best option. By nature, the assumption is that a cache can be invalidated at a predetermined time or even at a random time. While it is possible to associate persistent disk storage with a caching mechanism, it's not normally a requirement. A typical use case for caching is to store temporary data as a convenience for another consuming component. Relying on an inherently transient data store does not allow for patterns, such as **guaranteed delivery**, to be completely enforceable.

Again, how the mechanism is applied will be driven by the needs of the application and the business specifications behind it. Some popular caching technologies are as follows:

- Redis

- Memcached

- **Content delivery network** (**CDN**)

Stream-based technology

Stream-based brokers are, in many cases, like queue-based brokers. For example, there are generally places for messages to be sent (queues versus topics), some sort of computation or action to be taken, and a resulting message to be produced. A key difference between queue-based brokers and stream-based brokers is how the mechanism to receive, process, and retrieve data is handled. While queue-based technologies can be clustered to enable concurrency, stream-based systems provide concurrency and near-time visibility to data out of the box. This is facilitated by the lack of complex routing rules that can be found in a queue-based system. Stream-based systems allow raw information to be provided and the lack of potentially complex routing logic keeps touch time on messages very low. The following diagram shows the difference between a traditional queueing system and a streaming system in terms of concurrency, throughput, and resiliency:

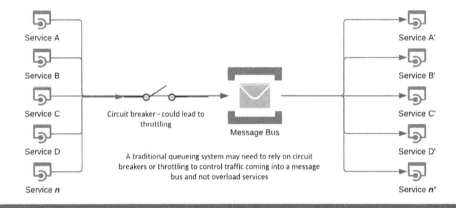

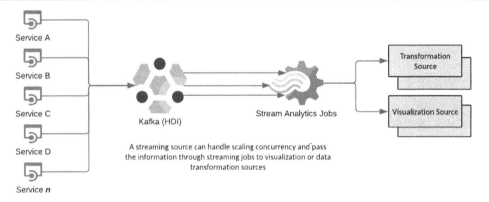

Figure 3.1 – Queue-based versus stream-based processing

This type of broker is intended to be used in systems that need to process high transaction volumes and varying amounts of data with little to no drag. Examples of stream-based brokers are as follows:

- Kafka
- EventStore

Having described the different types of message brokers, we will now turn our attention to how to use them – that is, through protocols, schemas, and delivery patterns.

Inspecting messaging protocols, schemas, and delivery patterns

Just as there are different types of message brokers, there are also different types of communication protocols that can be used, various schemas to define messages, and different types of message delivery. Choosing the option that best fits your application's needs is not always an easy task as more than one option may be applicable. As we examine the options in each category, we'll reflect on the choices that were made for the MTAEDA application and whether another option would be better suited.

Messaging protocols

Protocols related to programmatic communication are not necessarily exclusive to sequentially processed code or traditional client-server web applications. Different communication patterns can be leveraged across many types of software components. In the case of components that rely on messaging in some way, there are a few protocols that make more sense than others. Let's review those protocols in more detail.

AMQP

Advanced Message Queue Protocol (AMQP) is an open standard that was originally developed by John O'Hara that builds upon the simplicity of message queueing by focusing on the following principles:

- Message orientation
- Queueing
- Routing
- Reliability
- Security

AMQP is meant to provide a translatable protocol between various brokers and components that implement the pattern. The intent is to avoid or eliminate vendor lock-in with broker technologies or with broker-specific protocols.

MQTT

MQ Telemetry Transport (**MQTT**) is a newer protocol that is tailored toward **Internet of Things** (**IoT**) devices, specifically where message payloads need to be small and network connectivity could be severely limited or even non-existent at times. It also facilitates the use of TCP/IP not only as a delivery channel, but for sensor networks, Bluetooth, and more. Commercial and residential use cases can be addressed using MQTT. For example, many home automation hubs allow the use of an MQTT broker on a central device, with the ability to communicate over the aforementioned channels or even build workflows with visual tools such as **Node-RED**. This protocol could be considered for messages that are going to and from the turnstile and camera units, as per our example scenario.

Many other protocols can be used to send messages aside from AMQP or MQTT. Standard HTTP requests, **WebSockets**, and TCP socket connections can also act as conduits for transmitting messages from one place to another. Regardless of the method that's used to send the message, the content is ultimately what matters. Let's take a moment to look at how a message's content can be defined, as well as some pros and cons to each approach.

Standard and custom schemas

Using a pre-existing schema to define the structure of your messages can bring benefits and drawbacks, depending on the approach that's chosen. In some cases, having a format that is standardized and can be consumed by many different systems can be appealing. In others, having an internally defined schema to define messages may be easier to manage, especially if the messages are not meant to be consumed outside the application itself. Understanding the needs of your application, as well as the expected interaction patterns of applications that may be dependent on yours, will ultimately drive your decision. Let's take a moment and evaluate each option.

Using (open) schema standards

As we mentioned in *Chapter 1, The Sample Application*, we chose to use the **CloudEvents** open schema standard. The reasoning behind this choice was to enable events to be consumed natively within logging systems that support the schema, such as Azure Monitor, and to define a single pattern for message and event layouts that allows for easier developer consumption.

While this approach does bring consistency to all the events that are raised (and handled), there is the underlying issue of determining the data associated with each event. As the content of the message's object is serialized from domain events, entities, or handlers, the need to deserialize these messages may pose a bit of a challenge. If the target type cannot be loaded from a referenced library for any reason, this could cause the event handler to not function appropriately, thus adversely impacting the core functionality of the system.

Bring-your-own schema

It is entirely possible that keeping a lean (or implied) schema for each message could suit your needs. Even sending a serialized JSON instance of the object itself is a simple way to record what the content of that message is. If there are no additional system integrations to consider, this approach may be better.

A potential drawback to this approach would be *forward compatibility*. What if objects change over time? What if a new requirement is introduced to feed all the messages or processed messages into an aggregator? Small changes that may not seem like much at the time can pile up, making an integration event such as using an aggregator more painful. To help alleviate issues with forward compatibility, you can leverage a pattern such as **semantic versioning** to help set clear compatibility tracks, where major revisions imply breaking changes. Designing schemas that are rarely updated can also aid in avoiding forward compatibility problems.

Every design decision has potential upsides and pitfalls. Choosing the best option is ultimately up to you as you compose your application. How messages are delivered does not rest solely in your hands. This is where message delivery patterns become relevant.

Message delivery patterns

In general, when a message or event is produced, the expectation is that something, somewhere, will pick that up and perform an action based on its contents. Let's look at these different patterns and the expected behavior of each.

Guaranteed delivery

Guaranteed delivery is a concept that states that a message, once produced, will be delivered. Whether that's one copy or several, the scope of that delivery is only relevant to the notion that it will – and must – be delivered.

Exactly-once

The exactly-once pattern implies that a message or event will only be sent one time to one recipient. This implies that a durability mechanism needs to be in place not only on the producer side but on the consumer side as well. Many arguments have been made about the viability of this pattern since the controls that are needed to ensure compliance would need to be structured and potentially overbearing or difficult to maintain.

At-least-once

The at-least-once pattern is more traditionally known as a **broadcast pattern**, where at least one message or event will be sent, but it may be processed by many (or no) recipients. In the scope of using the producer-consumer pattern, this works well as many consumers may be interested in the produced event. Being able to write an event to a topic would, in theory, satisfy this pattern.

> **Important note**
>
> We will discuss different cloud-native resiliency patterns in *Chapter 4, Domain Model and Asynchronous Design*, when we dive into the domain code. To implement these patterns, we will be using Polly.net, an open source library meant to ease the adoption and implementation of these patterns. In situations where a component or a device in the field is unable to connect, employing the use of a resiliency pattern such as the **retry pattern** can help reduce the risk of messages not being processed, with the ultimate goal of persisting the message(s) locally if communication with the broker is not possible.

At-most-once

The at-most-once pattern is similar to the exactly-once pattern in that it intends to only allow one message to be processed. The major difference, though, is that the at-most-once pattern does not guarantee that anything will be processed. If a message is received, it will consume it no more than once. There is a broad implication of data loss with this pattern. Choosing to use this pattern comes with the burden of not only understanding the risk of data loss but also the responsibility of mitigating that risk.

Implementing message broker technologies

As we saw in the previous chapter, standing up and using a minimal Kafka instance can be done relatively quickly. This is great for localized testing; however, it does not translate into a production-grade infrastructure that's capable of handling the raw volume of events we may see with the application. While every configuration detail is not relevant to developing the domain code and the overall application, there are some points to keep in mind when you're setting up and configuring Kafka that can impact how software components may process events.

Now, let's walk through a high-level overview of the components that are needed to run Kafka, as well as relevant implementations and configurations that will enable resiliency and scalability.

Reviewing essential Kafka components

There are three primary components that you must have to establish a functioning Kafka instance. We've already talked about the broker, as well as topics. The final piece of the puzzle is the Zookeeper component. Zookeeper, according to the official `Apache Zookeeper` site, *is a centralized service for maintaining configuration information, naming, providing distributed synchronization, and providing group services*. In short, it minds the brokers in your Kafka instance and helps facilitate event routing, configuration updates, replication, and leader election. It also does the following:

- Event routing ensures events that are written to topics are written to the correct destination.

- When configuration updates are sent for Kafka brokers or other components, Zookeeper ensures those updates are applied uniformly.

- Zookeeper manages event replication across brokers and topic partitions based on the configuration that has been set.

- Zookeeper manages the primary active broker by declaring it as the lead node. This is known as leader election.

From an infrastructure perspective, this is something that requires a fair amount of planning and design. From a developer's perspective, it's good to understand what the base components are and how many brokers to use. For example, during local or integration testing, it may make sense to only use a single Kafka broker. In performance testing or production environments, the number of brokers would be increased significantly based on the anticipated volume of events moving through it.

Let's look at an example. Consider the following `docker compose` file:

```
---
   version: '2'
   services:
     zookeeper:
       image: bitnami/zookeeper
       hostname: zookeeper
       container_name: zookeeper
       ports:
```

```
      - "2181:2181"
    environment:
      ZOOKEEPER_CLIENT_PORT: 2181

  broker:
    image: bitnami/kafka
    hostname: broker
    container_name: broker
    depends_on:
      - zookeeper
    ports:
      - "29092:29092"
    environment:
      KAFKA_PORT: 9092
      KAFKA_ZOOKEEPER_CONNECT: 'zookeeper:2181'
      KAFKA_LISTENERS: EXTERNAL_SAME_HOST://:29092,
        INTERNAL://:9092
      KAFKA_ADVERTISED_LISTENERS: INTERNAL://broker:9092,
        EXTERNAL_SAME_HOST://localhost:29092
      KAFKA_LISTENER_SECURITY_PROTOCOL_MAP:
        INTERNAL:PLAINTEXT,EXTERNAL_SAME_HOST:PLAINTEXT
      KAFKA_INTER_BROKER_LISTENER_NAME: INTERNAL
      ALLOW_PLAINTEXT_LISTENER: "yes"
      JMX_PORT: 9999
```

This is a very basic Kafka setup with one broker and one Zookeeper node, along with some additional configuration on the broker node to allow internal traffic between Zookeeper and the broker, as well as external (localhost) traffic on port 29092. But what if we wanted to add another broker to the family? Using the existing configuration, you could update the compose file to look more like the following. The updates that have been made to the first broker node have been highlighted:

```
---
  version: '2'
  services:
    zookeeper:
      ...
```

```
broker1:
  image: bitnami/kafka
  hostname: broker1
  container_name: broker1
  depends_on:
    - zookeeper
  ports:
    - "29092:29092"
  environment:
    BROKER_ID: 1
    KAFKA_PORT: 9092
    KAFKA_ZOOKEEPER_CONNECT: 'zookeeper:2181'
    KAFKA_LISTENERS: EXTERNAL_SAME_HOST://:29092,
      INTERNAL://:9092
    KAFKA_ADVERTISED_LISTENERS: INTERNAL://broker1:9092
      ,EXTERNAL_SAME_HOST://localhost:29092
    KAFKA_LISTENER_SECURITY_PROTOCOL_MAP:
      INTERNAL:PLAINTEXT,EXTERNAL_SAME_HOST:PLAINTEXT
    KAFKA_INTER_BROKER_LISTENER_NAME: INTERNAL
    ALLOW_PLAINTEXT_LISTENER: "yes"
    JMX_PORT: 9999

broker2:
  image: bitnami/kafka
  hostname: broker2
  container_name: broker2
  depends_on:
    - zookeeper
  ports:
    - "29093:29093"
  environment:
    BROKER_ID: 2
    KAFKA_PORT: 9093
    KAFKA_ZOOKEEPER_CONNECT: 'zookeeper:2181'
    KAFKA_LISTENERS: EXTERNAL_SAME_HOST://:29093,
      INTERNAL://:9093
```

```
KAFKA_ADVERTISED_LISTENERS: INTERNAL://broker2:9093
  ,EXTERNAL_SAME_HOST://localhost:29093
KAFKA_LISTENER_SECURITY_PROTOCOL_MAP:
  INTERNAL:PLAINTEXT,EXTERNAL_SAME_HOST:PLAINTEXT
KAFKA_INTER_BROKER_LISTENER_NAME: INTERNAL
ALLOW_PLAINTEXT_LISTENER: "yes"
JMX_PORT: 10000
```

The Zookeeper configuration is not shown as it is identical to the first docker compose file. This setup will create a Kafka instance with two brokers and one Zookeeper node, which enables you to replicate topics between brokers if you wish. Take note of the ports on each of the brokers since they cannot use the same port for communication without creating collisions. Zookeeper will keep track of each broker in the cluster, so long as each has a unique port that it operates on.

While having more than one broker is interesting, interacting with the Kafka instance is done through the **command-line interface (CLI)**. To add a web-based UI administrative view, add the following service to the end of the docker compose file:

```
kafka-ui:
  image: obsidiandynamics/kafdrop
  ports:
    - 9100:9000
  environment:
    - KAFKA_BROKERCONNECT=broker1:9092
    - JVM_OPTS=-Xms32M -Xmx64M
  depends_on:
    - broker1
    - broker2
```

Kafdrop is an open source UI for Kafka that allows you to view the basics of your cluster, as well as create topics and view messages within those topics. The following screenshot shows the Kafdrop UI for Kafka:

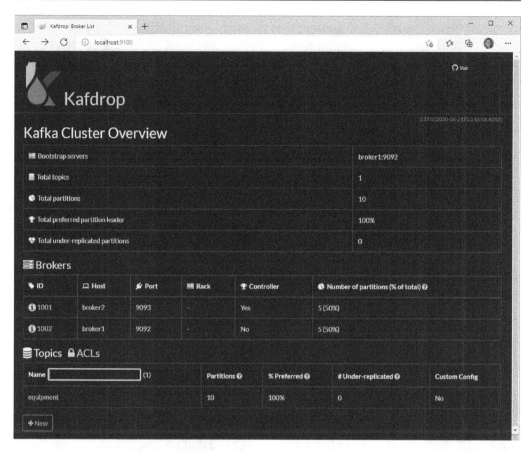

Figure 3.2 – The Kafdrop UI for Kafka

Enabling resiliency and scalability

As Kafka will be the main integration point for the producers and consumers within the MTAEDA application, it's important to consider resiliency and scalability. Resiliency, in this case, relates to the ability of Kafka to perform under duress – whether a broker is offline, communication between a producer or consumer takes longer than anticipated due to network load, or even Zookeeper being offline. In this context, scalability is related to how many brokers or Zookeeper nodes need to be available, as well as how many can be spun up to handle the additional load.

Three basic aspects of the Kafka cluster (that will enable resiliency) are the number of brokers that have been created, the replication factor for any topics you create, and the use of persistent storage for each broker to safeguard against data loss. We've already seen how to increase the number of brokers when using `docker compose`. When creating topics, you can specify the replication factor of the topic, which will default to the maximum number of brokers available. You don't necessarily need to use all the available brokers, but in the case of critical topics, it may make sense to utilize them all. The following screenshot shows the process of creating a new topic using Kafdrop, where the replication factor is specified:

Figure 3.3 – Creating a new topic in Kafka using Kafdrop

The last resiliency aspect is that of persistent storage. For our example, we would add a volume to the docker compose file that specifies a location on disk (locally) to map to when each broker is saving data. In some cases, persistence may not be as big of a deal. For instance, as part of the overall application, considerations may be made for data loss in the event of a regional outage. If the **recovery point objective** (**RPO**) is very low, such as minutes, more attention will be needed to ensure storage is fault-tolerant. If it is higher, such as hours or days, it may not be necessary to use fault-tolerant storage.

For scalability, using a docker compose file will not impact the ability to scale outside of adding more brokers manually to the file itself. Using technologies such as Kubernetes or Azure HD Insights, it is possible to configure autoscaling rules that will temporarily increase the number of nodes in the cluster to accommodate an increase in network traffic, CPU utilization, RAM utilization, and other metrics. Managed offerings such as Confluent Cloud will also allow you to specify scaling requirements.

Planning for scalability events, regardless of whether they are manual or automatic, is something you should consider when you're planning out the entire application from an infrastructure perspective. From a developer's perspective, this activity may likely be less relevant to daily tasks since the primary focus is being able to write consistent and durable code. It is generally a good practice to have a basic understanding of how scalability impacts the application and where it becomes relevant.

Summary

In this chapter, we took a closer look at message brokers and related topics. From defining what a message broker does, different communication protocols, the message schema, delivery patterns, and even implementation, we've laid the foundation for a better understanding of how messages and events move through the system. While it's not essential to understand every detail related to these topics, having awareness of these items will help tailor your understanding of messaging operations, how the producer-consumer pattern leverages these operations, and how best to code for (and anticipate) situations related to successful and unsuccessful message processing.

In *Chapter 4, Domain Model and Asynchronous Design*, this inherent knowledge will be further reinforced as we analyze the domains and the associated code bases for each. In addition, we will introduce resiliency patterns, which help safeguard against potential data loss or issues with overwhelming producers or consumers in each domain.

Questions

Answer the following questions to test your knowledge of this chapter:

1. What are the three main types of message brokers?
2. Is the use of AMQP predominant in the messaging world?
3. Can HTTP be used as a transport protocol for sending and receiving messages?
4. Which is better for defining a message schema for an application – a standardized format or a single format per message type?
5. Of the different message delivery patterns we discussed, which pattern is most likely to result in unintentional data loss?
6. Is it possible to have a functioning Kafka cluster with only one broker? Why or why not?
7. What is the importance of Zookeeper in the Kafka cluster?
8. Is it only possible to create topics via the command line? Why or why not?

Further reading

To learn more about the topics that were covered in this chapter, take a look at the following resources:

- *RabbitMQ vs Kafka: Comparing Two Popular Message Brokers*, by *SPEC India*: `https://www.spec-india.com/blog/rabbitmq-vs-kafka`

- *Getting Started with EventStore*: `https://developers.eventstore.com/server/v21.6/docs/introduction/#getting-started`

- *Home | AMQP*, by The *AMQP Foundation*: `https://amqp.org`

- *MQTT: The Standard for IoT Messaging*, by *MQTT*: `https://mqtt.org`

- *Exactly Once Delivery*, by *Cloud Computing Patterns*: `https://www.cloudcomputingpatterns.org/exactly_once_delivery/`

- *At Least Once Delivery*, by *Cloud Computing Patterns*: `https://www.cloudcomputingpatterns.org/at_least_once_delivery/`

- *CloudEvents: A specification for describing event data in a common way*: `https://cloudevents.io/`

- *Apache Zookeeper*, by *The Apache Foundation*: `https://zookeeper.apache.org/`

- *Kafdrop*, by *Obsidian Dynamics*: `https://github.com/obsidiandynamics/kafdrop`

4

Domain Model and Asynchronous Events

When you're deciding on a communication pattern for an application, it is normal to look at the circumstances under which components may communicate with one another. Typical web applications follow a request/response pattern, where a request is made to the server and a response is returned to the client and handled appropriately. For applications looking to handle high throughput in those communication channels, a standard request/response pattern is not desired as each request would expect and wait for a response, leading to latency for requests that have been submitted while one is still processing. With hundreds of thousands of messages being sent per second, the application would be slowed to a halt as every request is met with a corresponding response. Scenarios such as this call for a more resilient and asynchronous messaging pattern.

This chapter intends to dive deeper into the domain model for the sample application. To do so, we will cover the following topics:

- Reviewing domain commands and events
- Using asynchronous actions
- Exploring the new async features of .NET 6

By the end of this chapter, we will be able to do the following:

- Understand the application's domain model in further detail, including commands, events, event handlers, entities, and value objects.

- Understand why asynchronous design is beneficial and how its implementation is critical to enabling applications to use **Event-Driven Architecture** (**EDA**).

Technical requirements

You can find all the code examples for this chapter in this book's GitHub repository at `https://github.com/PacktPublishing/Implementing-Event-Driven-Microservices-Architecture-in-.NET-7/tree/main/chapter04` and `https://github.com/PacktPublishing/Implementing-Event-Driven-Microservices-Architecture-in-.NET-7/tree/main/src`.

> **Important note**
> The links to all the white papers and other sources mentioned in this chapter can be found in the *Further reading* section at the end of this chapter.

Solution structure

The structure of the overall MTAEDA solution is broken up into two major divisions:

- The core libraries that provide base functionality for use within all domains
- The domains themselves, each isolated from the other using separate solution files

The core libraries are set up to reduce complexity across the various domain solutions and are intended to offer very base functionality that can be leveraged by each domain. This functionality includes interfaces for domain objects, interfaces and classes for interaction with Kafka, and utility classes to facilitate packing and unpacking domain events.

Each domain solution follows a standard format, including projects that address the following areas:

- Domain logic
- Infrastructure (data access and repositories)
- Command-based API endpoints (creating, changing, or updating information)
- Query-based API endpoints (retrieving information only)

- Listener services (proxies for event handlers)
- Test projects

Using a standard layout for all your solutions ensures that locating specific projects, files, or references is easier, regardless of what solution you are in. The only exception to that layout is the core library, which contains important shared objects and interfaces. The core library contains custom code, as well as references to external libraries, such as **MediatR**. **MediatR** is an implementation of the **mediator pattern**, which is meant for event-driven architectures and domain-driven design. Let's explore the core library components in a bit more detail.

Core library review

The core library for the MTAEDA application consists of interfaces and concrete implementations that all domains can leverage. These fall into domain interfaces, infrastructure interfaces and classes, and utility classes. There are several domain interfaces, as illustrated in the following table:

Name	Description
`IDomainAggregateRoot`	Allows domain aggregate objects to be identified (decorator only).
`IDomainEntity`	Allows domain entities to be identified and requires an implementation of a string field for domain data.
`IDomainEvent`	Implements `INotification` from the MediatR library. Allows domain events to be identified and requires an implementation of a string field for domain data.
`IDomainEventHandler`	Implements `INotificationHandler<T>` from the MediatR library, where `T` is an `INotification` object. Allows event handlers to be identified and requires the `Handle()` method to be implemented.
`IDomainValueObject`	Allows value objects to be identified within the domain (decorator only).
`IRepository<T>`	The generic `T` type must be an `IDomainEntity` object. Allows a base set of methods to be enforced for any repository implementation.

Table 4.1 – Domain interfaces

In terms of the infrastructure, two interfaces and two concrete Kafka provider implementations are supplied to offer support for event sending and receiving. Normally, these provider implementations would be in each domain's infrastructure project. For the sake of simplicity, however, they have been placed in the core library for wider use.

The following table illustrates the components that are available under the infrastructure folder:

Name	Description
`IEventReaderProvider`	Interface for read operations around events.
`IEventWriterProvider`	Interface for write operations around events.
`KafkaConsumer`	Implements `IeventReaderProvider`. A provider for connecting to a Kafka instance and consuming events from one or more topics.
`KafkaProducer`	Implements `IeventWriterProvider`. A provider for connecting to a Kafka instance and publishing events to a topic.

Table 4.2 – Infrastructure interfaces and classes

As far as utility classes go, only one is provided by the core library. The `EventUtil` class provides a method for packing and unpacking domain events, event handlers, and exceptions. This facilitates the use of the **CloudEvents** schema, as discussed in *Chapter 1, The Sample Application*, for the standard format of all events, but it still provides a way to get to the original object and perform any additional operations.

> **Important note**
> The following legend applies to all the diagrams in this chapter.

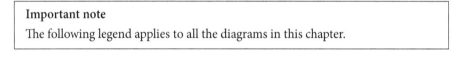

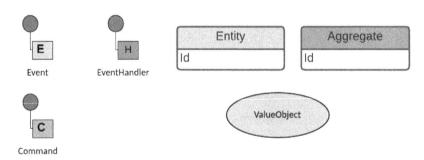

Figure 4.1 – The legend followed for all the diagrams

Now that we understand the common structure that's used for each domain, let's drill into the model of each domain. We will review not only every domain's purpose but also the core components within each domain and how they provide value to the overall application.

Reviewing domain structures and components

In *Chapter 1, The Sample Application*, we took a quick look at the outlined domains for the application, as well as a few (but not all) of the commands, events, entities, and other domain objects. Now that we are armed with some knowledge of how the consumer-producer pattern works and how the message broker facilitates that pattern, we will dive into each domain at length to review the pertinent objects within them.

Equipment

The equipment domain is of critical importance. Without a means to manage events that are related to the turnstile units, as well as the cameras in each unit, the application itself does not serve much of a purpose. The following diagram shows the domain architecture for the equipment domain:

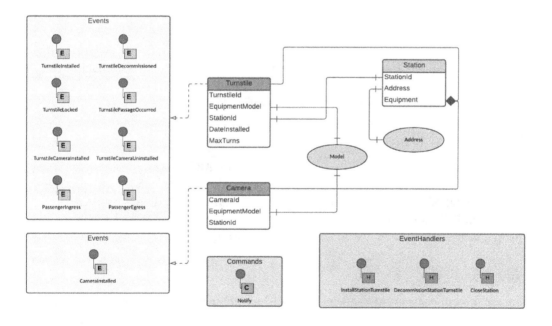

Figure 4.2 – The domain architecture for the equipment domain

The equipment domain is central to the application. Many events are triggered by events that originate from this domain.

Aggregates

The equipment domain leverages two aggregate root objects to manage all interactions within the domain:

- **Camera**: This is used to track basic device information related to the camera and associated maintenance events. It includes the camera ID, the equipment model information, and the station ID.

- **Turnstile**: This is used to track basic device information related to the turnstile and associated maintenance events. It includes the turnstile ID, the equipment model information, the maximum number of turns before maintenance is required, the install date, and the station ID.

You will notice that the **camera** and **turnstile** aggregates appear in other domains as well. In some cases, they have different information associated with them. Since these are objects that are relevant across several domains, it is common to see this level of duplication. The events and entities they interact with, however, will be relegated to that particular domain.

Entities

The equipment domain interacts with the station domain entity to help track equipment. This is a domain entity that contains minimal station information (ID and address), as well as a full inventory of equipment at that location.

Events

There are nine core events within the equipment domain. All but one of these events is related to actions that can occur with a **turnstile**. The only event associated with a **camera** is **CameraInstalled**, which is triggered after **TurnstileCameraInstalled** is handled. **CameraInstalled** intends to pass the **model** information of the camera back to the domain and associate it with the **turnstile** and the **station**.

Event handlers

There are three specific event handlers, two of which are related to the turnstiles located within a station (installation and decommissioning). There is also an event handler for closing stations, which would lead to decommissioning all the equipment objects related to that station.

Value objects

For value objects, we have the following entities:

- **Address**: A standard address format that's used to locate the station.

- **Model**: Equipment model information related to a turnstile, camera, or another type of station equipment.

Now that we have a better understanding of the workings of the equipment domain, let's move on to the station domain.

Station

The station domain deals with domain objects directly related to physical mass transit stations. This is more of a supportive domain in the context of the application as the primary interactions are with the equipment at each station. There is a need to identify where each piece of equipment is, and some core data around that, which is where the station domain comes in. The following diagram shows the domain architecture for the station domain:

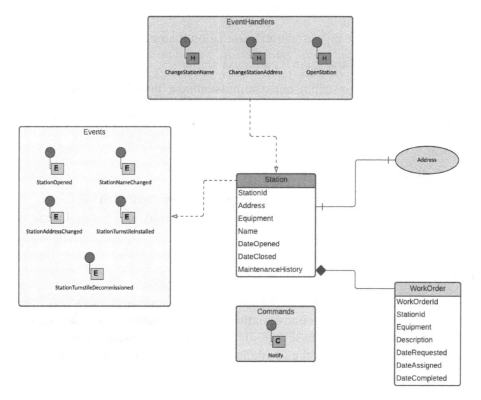

Figure 4.3 – The domain architecture for the station domain

While it may seem that this domain is rather small, it does serve a greater purpose, giving a location to equipment, maintenance records, scheduling, and identifying potentially dangerous fugitives at large.

Aggregates

The primary aggregate for this domain is the **station**. This aggregate tracks basic information about the station, including whether it is still open (operational) or closed. It also allows the history of work orders associated with the station to be pulled.

Entities

There is only one domain entity associated with the station, and that is **WorkOrder**. This is a domain entity that contains information about a specific work order related to a maintenance schedule. It can apply to any type of equipment and captures when the work order was created, when it was assigned, and when it was completed.

Events

There are six domain events related to stations:

- Two are related to a change in station information (name or address).
- Two events indicate a change in the station's status.
- Two events deal with installing or decommissioning a turnstile in a station.

These events relate directly to the event handlers for the domain, listed in the next paragraph.

Event handlers

The station domain has three event handlers to manage events related to the station, namely a change in name, address, or status. Status indicators for a station include active, closed, and offline.

Maintenance

The maintenance domain deals with events related to any maintenance that's needed on turnstiles and cameras at any mass transit station. While not responsible for scheduling maintenance, it is responsible for raising events that state that maintenance is needed, which, in turn, will raise events in the scheduling domain to initiate the creation of a work order. This is shown in the following diagram:

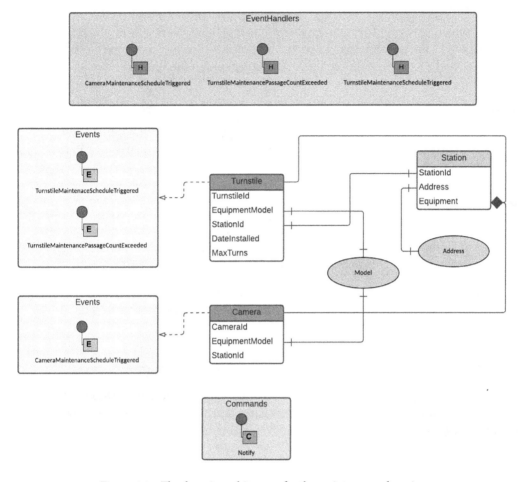

Figure 4.4 – The domain architecture for the maintenance domain

The maintenance domain is managed by several different domain objects. Let's take a look.

Aggregates

The maintenance domain leverages the **camera** and the **turnstile** aggregates, as shown in the equipment domain, though the events that each control are different. In the case of this domain, the events that are leveraged only reference the events that can be triggered to invoke maintenance requests.

Entities

The sole domain entity for the maintenance domain is the **station**, a minimalist domain entity that tracks the equipment for a particular station. This only contains the station ID, the address, and a list of equipment that's been registered to the station.

Events

Two events indicate when a maintenance request is triggered, one for a camera and one for a turnstile. There is also an event that can be raised when a turnstile exceeds its maximum number of turns.

Event handlers

There are two event handlers for managing an incoming maintenance request (one for turnstiles and one for cameras) and one for managing an incoming event that's triggered by the turnstile exceeding its maximum turn count.

With that, we have seen how the maintenance domain handles requests for maintenance activities to be performed. Next, we will look into the scheduling domain and its relationship with the maintenance domain.

Scheduling

The scheduling domain deals with adding, updating, or removing scheduled maintenance events. These changes in schedule status are triggered primarily from the maintenance domain but can be triggered from the equipment domain as well in the case of a jammed turnstile:

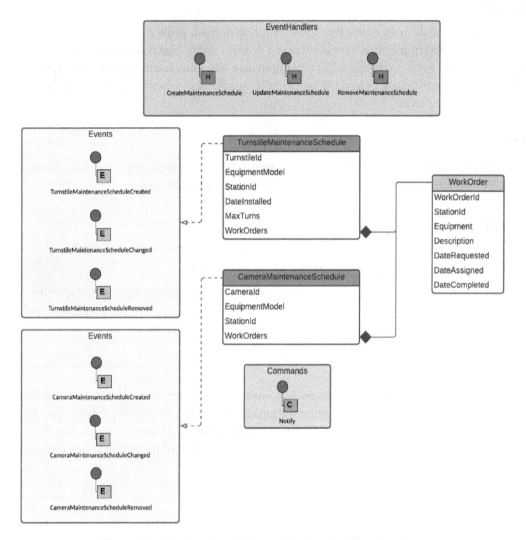

Figure 4.5 – The domain architecture for the scheduling domain

The scheduling domain is split into the object types shown in the preceding diagram, which are responsible for scheduling maintenance activities. Let's take a closer look.

Aggregates

The scheduling domain has two aggregate roots that manage scheduling maintenance activities. One is **CameraMaintenceSchedule**, which deals with any maintenance schedule records related to camera equipment. The other is **TurnstileMaintenanceSchedule**, which deals with any maintenance schedule records related to turnstiles.

Entities

There is only one domain entity that the scheduling domain deals directly with, and that is the **WorkOrder** entity. **WorkOrder** is a domain entity that contains information about a specific work order related to a maintenance schedule. It can apply to any type of equipment and captures when the work order was created, when it was assigned, and when it was completed.

Events

There are a total of six events in the domain, split across the two aggregates evenly. There are create, change, and remove events for turnstiles as well as cameras.

Event handlers

Three event handlers exist within the domain, one for each action type that can be performed on a schedule: creation, modification, or removal.

As we've seen, the scheduling domain manages how work orders are created to facilitate maintenance activities taking place on equipment located within a station. Next, we'll look at the notifications domain, which is a crucial part of the application.

Notifications

The notifications domain deals with sending notifications to various consumers as a direct result of an event being handled. Configurations are stored in the domain to help classify the type of notification, where it should go, and what template to use when deriving the output. The following diagram shows the domain architecture of the notifications domain:

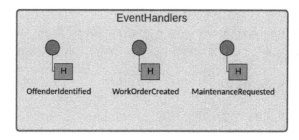

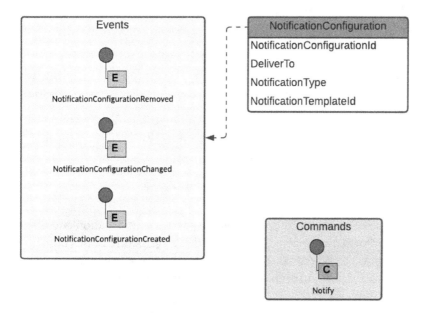

Figure 4.6 – The domain architecture for the notifications domain

Notifications are a critical part of the application since many domains need to send notifications out once certain events have been handled. Let's take a look at the notifications domain in a bit more detail.

Aggregates

There is only one aggregate root within the notifications domain, and that is **NotificationConfiguration**. This aggregate contains information about a specific notification configuration setup. It houses the configuration identifier, the address to deliver the notification to, the type of notification, and a template identifier that can optionally be used to pull in a template to format the body of the notification itself.

Events

This domain has three events related to **NotificationConfiguration** – one for creating a configuration, one for updating it, and one for removing it from use.

Event handlers

There are three event handlers:

- **MaintenanceRequested**: This will handle a maintenance request being raised and create a **WorkOrder** to initiate the process.

- **WorkOrderCreated**: This will handle the initial processing of **WorkOrder** once the maintenance request event has been handled.

- **OffenderIdentified**: This will send a notification to the appropriate recipients when an event is raised by the identification domain stating that a match has been found in the offender list.

Now that we've covered the notifications domain, we can move on to the passenger domain.

Passenger

The passenger domain is reserved for future use, where payment information and other program membership information may be stored. The overall structure of this domain has been planned out, but the code for any implementations has not been written. The following diagram shows the domain architecture of the passenger domain:

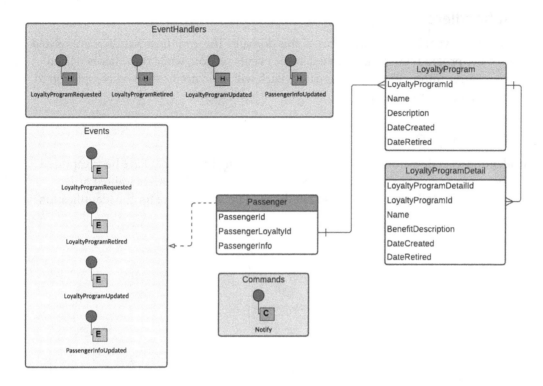

Figure 4.7 – The domain architecture for the passenger domain

While this domain is not widely used across the application at this point, it does have a tie-in to the identification domain and allows for future expansion into more customer-facing interactions.

Aggregates

The passenger domain has only one aggregate root, which happens to be the **passenger**. This aggregate contains the identifier for the passenger along with a reference to the **LoyaltyProgram** entity. A separate structure for passenger information (**PassengerInfo**) is also used to update basic information for the passenger.

Events

There are three events related to the loyalty program itself: one for requesting a new program for the passenger, one for updating the information, and one for retiring or closing out the program. There is also an event for updating the passenger's information by way of the **PassengerInfo** object.

Event handlers

A total of four event handlers are a part of this domain. Three of those handlers are related to the loyalty program events mentioned in the events section, while the other is related to the passenger information being updated. Each will fire a notification on completion to inform the passenger of changes.

Identification

The identification domain is solely responsible for taking images that have been captured by station cameras and validating them against a known list of fugitives wanted by law enforcement. The following diagram shows the domain architecture for the identification domain:

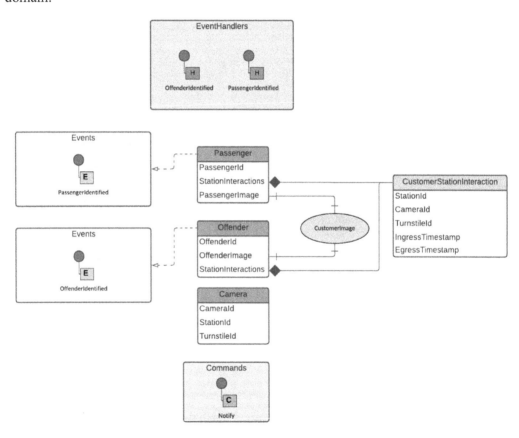

Figure 4.8 – The domain architecture for the identification domain

The identification domain serves a very focused purpose – allowing potentially dangerous fugitives to be recognized and alerting the appropriate personnel for additional investigation.

Aggregates

There are three aggregates:

- **Camera**: This is used to track basic device information related to the camera that has made the identification. It will only store the camera identifier, station identifier, and turnstile identifier.

- **Offender**: This is used to track the offender identifier, any station interactions (ingress and egress), and the offender's image.

- **Passenger**: This is used to track the passenger identifier, any station interactions (ingress and egress), and the passenger's image.

While there are three aggregates within the domain, there is only one entity explicitly defined, as we will see next.

Entities

The only domain entity the identification domain has is the **CustomerStationInteraction** entity. This records information about specific interactions at a station, such as the turnstile the passenger used, as well as the ingress (entry) or egress (exit) timestamp. While both timestamps may have values for a single station, it is more common that one or the other will have a value based on the passenger's travel.

Events

The passenger domain only produces two events. One is for passenger identification, while the other is for offender identification. The offender identification event is raised when the camera interacts with facial recognition services to identify a fugitive on a selected list.

Event handlers

Just as the passenger domain only has two events, it also only has two event handlers. Each event handler matches up with either the passenger or offender identification. Each event handler will create a record of the identification using the **CustomerStationInteraction** entity, and in the case of an offender, a notification will also be sent.

> **Important note**
> Please take some time to review the source code for each domain solution, as well as the shared code library (MTAEDA.sln), in the src folder of the main repository.

With that, we've looked through the details of each domain, understanding how they operate as well as how they interact with each other's events. As we have learned throughout this book, these events do not require immediate responses. Now, let's look at asynchronous actions.

Using asynchronous actions

There are plenty of use cases for implementing synchronous actions within an application. At some point, those synchronous actions will not be able to scale out should the load on the application increase. This is where asynchronous actions come in. These actions allow many executions to be made using different contexts, whereas synchronous actions will block other actions from executing until the invoked action is complete. This ability to execute without the need to wait for a response allows large-scale message sending without any concern of thread-blocking, which can lead to longer wait times and less than desirable performance.

The notion of asynchronous actions has been a paradigm in programming for many years. Using callback functions, which are meant to be executed when one method has completed and returned a value, has been a construct in JavaScript, as well as other languages. Specific patterns such as asynchronous JavaScript and XML (**AJAX**) were some of the building blocks of web applications using frameworks such as **jQuery**. Later on, constructs such as **promises** also helped to extend asynchronous method calls through inline function handlers, providing a more succinct way to define callbacks.

In languages that utilize the **.NET Framework**, functions can be assigned to event handlers or to **delegates**, which will execute when a method or process triggers them synchronously. There is also a more structured pattern known as **Task Asynchronous Programming** (**TAP**), which leverages the `Task` and `Task<T>` objects in the system threading namespace to perform asynchronous operations. More information about TAP can be found in the *Further reading* section at the end of this chapter.

Having covered the basics of asynchronous actions and how they can be used to bolster resiliency and throughput, let's look at some of the additional benefits of using the asynchronous programming model.

Benefits of asynchronous programming

Primarily, asynchronous programming allows you to execute a greater sum of code than synchronous programming. It also allows you to leverage compute resources differently by only using resources while a method is executing. This avoids unnecessarily assigning compute resources and should lead to more consistent response times overall. This allows you to execute more complex tasks and leverage concurrency while not needing to worry about thread pool management.

Asynchronous methods are more efficient in how they utilize available resources. Longer-running operations can use the `yield` construct in C# to continue executing while other operations can be invoked as there are still resources available to them. When coupled with concepts such as parallel execution, asynchronous execution can facilitate further scaling of the application while preserving its responsiveness.

Asynchronous parallel processing

In .NET 6, a new construct was introduced known as `Parallel.ForEachAsync`. Those familiar with the notion of parallel execution will likely have some knowledge of the `Parallel.ForEach` method. This new asynchronous method allows you to perform that parallel execution while also enabling asynchronous execution.

Let's take a look at an example of implementing `Parallel.ForEachAsync` in a console app that has multiple consumers pointed at different topics in Kafka. In this case, we'll be consuming three different topics using the aforementioned method, allowing all of them to run in parallel and consume events as they are available.

We will start by using a top-level program to pull in configuration information from an `appsettings.json` file and create a consumer for each topic name listed in the configuration. The `appsettings.json` file is parsed by making a call to the `ConfigurationBuilder` object to retrieve the information. The list of topics is stored as a pipe-delimited string, and after splitting that string, an initial loop is executed to set up each consumer and add it to a list of consumer objects. The code for setting up each consumer is as follows:

```
var consumer = new ConsumerBuilder<int, string>(
   new ConsumerConfig()
   {
       GroupId = config["ConsumerConfig:GroupId"],
       BootstrapServers =
          config["ConsumerConfig:BootstrapServers"],
       AllowAutoCreateTopics = true,

   }).Build();
consumer.Subscribe(topic);
consumerList.Add(consumer);
```

We'll also set up the body of the `Parallel.ForEachAsync` loop to take in the list of consumers, an object specifying parallel options, and a function that will run asynchronously, operating on the consumer as well as passing in `CancellationToken` to allow a consumer to be shut down, should a cancellation request get sent in:

```
await Parallel.ForEachAsync(consumerList, options, async (s, t)
  =>
{
    if (t.IsCancellationRequested)
    {
        s.Close();
    }
    else
    {
        await Task.Factory.StartNew(() =>
        {
            Console.WriteLine($"Consuming
              {s.Subscription.First()}...");
            s.Consume(t);
        });
    }
});
```

When you're debugging the code, please ensure you are also running the Kafka instance by running `docker compose` up from the code folder from *Chapter 3*, *Message Brokers*. Ideally, you could add code to invoke event handlers for each of the topics that are being consumed. Try experimenting with adding more topics to the list or allowing the code block to resolve an event handler based on the event's metadata.

We have covered a lot of ground regarding not only the overall domain model as well as how to take advantage of parallel execution with asynchronous actions. As we continue to work on the sample application, features such as these will become more prevalent, and understanding how and when to leverage them will be beneficial.

Summary

In this chapter, we took a much closer look at each domain in the MTAEDA application. This included a deep dive into each domain's aggregate roots, events, event handlers, entities, and more. Having this more comprehensive understanding of each domain helps not only to drive understanding of how the application components work together but also how topical events can be consumed by more than one domain.

We also investigated how to use asynchronous operations. We also learned about the `Parallel.ForEachAsync` method, and how to implement it. Using the example code, you should be able to experiment with different configurations and execution patterns.

In the next chapter, we will dig into the details of setting up the full local development environment for the application. This includes configuring your IDE as well as ensuring there is test infrastructure available locally to test with.

Questions

Answer the following questions to test your knowledge of this chapter:

1. Which domain is responsible for managing details around work orders that have been created to perform maintenance?
2. What is the primary reason for using the mediator pattern, and how does the sample application implement this?
3. Does asynchronous programming imply multithreaded execution?

Further reading

To learn more about the topics that were covered in this chapter, take a look at the following resources:

- *MediatR GitHub Repository*, by Jimmy Bogard, available at `https://github.com/jbogard/MediatR`.
- *MediatR Wiki*, by Jimmy Bogard, available at `https://github.com/jbogard/MediatR/wiki`.
- *What's New in C#11*, available at `https://docs.microsoft.com/en-us/dotnet/csharp/whats-new/csharp-11`.
- *Asynchronous Programming*, available at `https://docs.microsoft.com/en-us/dotnet/csharp/async`.
- *AJAX Introduction*, by W3Schools, available at `https://www.w3schools.com/xml/ajax_intro.asp`.
- *Delegates (C# Programming Guide)*, by Microsoft, available at `https://docs.microsoft.com/en-us/dotnet/csharp/programming-guide/delegates/`.

- *The Task Asynchronous Programming model*, available at `https://docs.microsoft.com/en-us/dotnet/csharp/programming-guide/concepts/async/task-asynchronous-programming-model`.

- *Parallel.ForEachAsync in .NET 6*, by Scott Hanselman, available at `https://www.hanselman.com/blog/parallelforeachasync-in-net-6`.

- *Parallel ForEachAsync method*, available at `https://docs.microsoft.com/en-us/dotnet/api/system.threading.tasks.parallel.foreachasync?view=net-6.0`.

Part 2: Testing and Deploying Microservices

This part will review the various components of local environment setup, containerization of code, testing, deployment, and observability of microservices utilizing an EDA approach.

This part has the following chapters:

- *Chapter 5, Containerization and Local Environment Setup*
- *Chapter 6, Localized Testing and Debugging of Microservices*
- *Chapter 7, Microservice Observability*
- *Chapter 8, CI/CD Pipelines and Integrated Testing*
- *Chapter 9, Fault Injection and Chaos Testing*

5
Containerization and Local Environment Setup

So far, we have seen that our modular solution design, with services implementing the **Single Responsibility Principle (SRP)**, can quickly lead to a wealth of running components within our application. Working on these as a single contributor could be overwhelming, and working on isolated domain teams may lead to hiccups when performing integration testing. Establishing a pattern that will address the potential for issues during integration, along with a lower barrier to entry for developers, can pay huge dividends down the road. Throughout this chapter, we will be doing just that—setting up a pattern for usage and deployment that will carry on across developer environments and, ultimately, to production.

Throughout this chapter, we will be doing the following:

- Reviewing containerization fundamentals
- Setting up the local environment
- Using Dockerfiles to build and run locally

By the end of this chapter, you will be able to do the following:

- Explain how the isolation of functionality enables further development, testing, and deployment.

- Set up your local development environment to allow for building Docker images and running Docker containers.

- Create a Dockerfile, identify the use of sequential execution versus multi-stage execution, and build applications during the image creation process.

Technical requirements

You will find all the code examples for this chapter in the folder for this chapter on GitHub here: `https://github.com/PacktPublishing/Implementing-Event-driven-Microservices-Architecture-in-.NET-7/tree/main/src`.

> **Important note**
> The links to all the white papers and other sources mentioned in the chapter are provided in the *Further reading* section towards the end of the chapter.

Reviewing containerization fundamentals

While it is assumed that you have some experience with Docker already, it's worth the time to quickly review some of the fundamentals of working with containers and some of the inherent benefits of using them. We will first cover the fundamentals of isolation, portability, reusability, layering, and abstraction.

Regarding **isolation**, **Docker** allows focusing on one process within the confines of a container. In the world of Docker, containers are immutable and are created from images that can be utilized directly from the Docker registry or can be used from a private registry for images that are custom built for specific needs. This notion of **immutability** allows for isolation to be confined to just the process that the container needs and not to any outside influences or mutations. That's not to say that outside influences cannot modify things within a Docker container. The immutability comes in when a container is restarted, and whatever information was modified during that attachment or connection is deleted.

The **portability** aspect of containers allows consistent execution across many different environments. For example, running a container on your local development machine versus running a container in a Docker instance or a cluster will all yield the same result. Some nuances at the operating system layer will drive certain decisions, such as programs that are inherently dependent on Windows as an operating system. One possible example of a Windows-based dependency might be a dependency on Win32 DLLs for a specific program. For language-based services and executables, if the language is truly cross-platform (.NET, Python, Java, and so on), it will not matter what operating system the container is running on.

The notion of **reusability** does intertwine a bit with that of layering. The idea is that one image can spawn N number of containers based on certain factors, such as the orchestration engine being used, resources within that orchestration engine, and overall system need. Therefore, one image is all that's truly required in order to spin up multiple containers running the same executable or service. Furthermore, when images are built, there is a notion of reusability with how Docker will store updates to the image. Docker uses a file system based on layers identified by checksums, reducing the need to create unique file system restore points based on common operations. When viewed in sequence, the file system appears to have different layers.

As an example, one of the ways in which Docker optimizes the file system is that for base images, using a base operating system of Ubuntu, for example, all of the bare minimum requirements to ensure that the container will run are already installed and captured as checksums. When you go to build a new image based on one of the preexisting base images, those checksums will come over as a part of that base image, eliminating the need for you to install bare requirements for operation. Now, you will need to install any specifics for your application based on language or other features. Each line that gets added to the configuration will create a new checksum and, subsequently, a new file system checkpoint.

Abstraction using containers is another area that can fundamentally change how you write software components. Using containers, you can abstract away a good amount of technical complexity from the primary concern of running the service or program you wish to run. This includes worrying about the underlying operating system, other items installed on the same host, and even security. In fact, it's considered best practice to limit the scope of the identity running the container to safeguard against potential attacks using elevated permissions. In many cases, base Docker images will default to the root account for execution, which provides far too much control. Creating a scoped service principal or user to use when executing the container helps safeguard against such attacks.

These topics in and of themselves can be massive and are far beyond the scope of this book. However, understanding how these fundamentals can benefit you as a developer and enable you to focus more on the pieces you absolutely need and less on other ancillary operational bits you may not.

Development benefits

The portability and reusability fundamentals are rather important when it comes to local development. Whether using Docker Compose, minikube, or another way of locally running Docker containers, the process that executes within that container will be consistent across any of those methods. This also means that developers working on different operating systems can technically run the same containers, thus enabling a seamless experience regardless of the developer's operating system of choice.

One of the ways in which using containers is beneficial to the development life cycle is that there is now an ample number of tools that allow you to debug containers without having to leave your desktop. Over the past several years, tooling from Docker, along with Microsoft and other open source partners, has enabled **integrated development environments** (IDEs) to be more efficient in helping build and test and debug container-based applications. This tooling reduces friction for developers, as debugging will just seem natural as opposed to needing to connect two different processes or command-line utilities to debug a specific function or piece of code.

Another benefit to the development life cycle can be seen in teams who are working with other domain teams to build a larger platform or application. To perform unit and integration tests locally to validate changes made in an isolated domain, pulling down images from other domain teams and then creating containers from those images allow for an easier way to leverage stable code to test against.

Having reviewed some of the principles and benefits of using containerization in the software development life cycle, we can now look to enabling our development environment to take advantage of these benefits.

Setting up the local environment

The developer environment is a critical aspect of enabling nimble iterations, shortening the feedback loop, and encapsulating security best practices during the development life cycle. Traditionally, a developer may have set up all of the dependencies, preferred libraries, and command-line utilities, as well as the IDE, on a physical workstation. While that can be done, advances in technology (also leveraging containerization principles) also enable modern developers to take advantage of reusable environments hosted in the cloud.

This section will look at two scenarios aimed at a repeatable and reliable developer setup. We will first look at configuring a local workstation environment, then move on to examine the world of virtual developer environments through utilities such as **GitHub Codespaces** and **Dev Containers** for **Visual Studio Code**.

Creating local infrastructure using Docker

As we've seen in previous chapters, using Docker and Docker Compose can speed up your local development time by providing you with a local representation of the infrastructure you might see in the cloud. But what about different implementations, such as a Kubernetes cluster? Let's look first at running the stack locally with Docker Compose, and then we will shift to utilizing Kubernetes as the target for infrastructure.

Standing up infrastructure using Docker Compose

One nice thing about the work we have done previously is that it is completely reusable in our scenario now. To get started, copy over the `docker-compose.yml` file from the source files in the `chapter03` folder to any of the domain solution folders in the `src` directory. This will provide you with a baseline Kafka implementation that you can connect to when debugging service code.

For the purpose of this walkthrough, we will be using the `MTAEDA.Equipment` solution to illustrate where to place the file, and **Visual Studio** to illustrate how to trigger the startup of the Docker services.

To start, your domain solution root folder should look like *Figure 5.1*, where the `docker-compose.yml` file has been copied in:

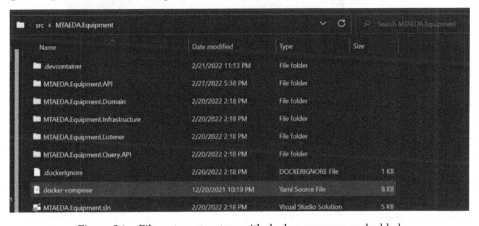

Figure 5.1 – File system structure with docker-compose.yml added

If you're using Visual Studio and have the Docker tools installed, you can right-click on a project that does not currently have a Dockerfile associated with it, and through the context menu, go to **Add | Docker Support…** to enable your project. You'll notice that the generated Dockerfile contains several FROM directives within it, which is indicative of a concept known as multi-stage Dockerfiles. We'll examine multi-stage Dockerfiles later in this chapter. As a reference, *Figure 5.2* illustrates where the **Docker Support…** menu item can be found:

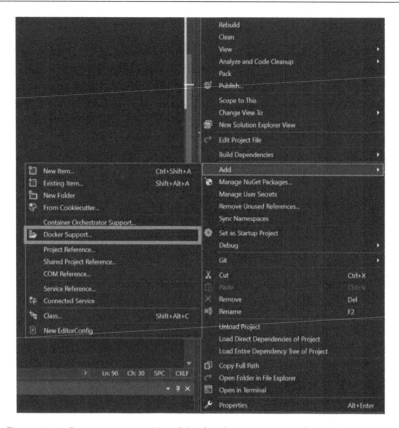

Figure 5.2 – Context menu in Visual Studio showing options for Docker support

Immediately above the **Docker Support…** menu item is another menu item for **Container Orchestration Support….** This will facilitate adding orchestration support for Docker Compose, but that does not mean you cannot add support for other orchestration engines. We will take a moment now to explore integrating with one of those orchestration engines—**Kubernetes**.

Standing up infrastructure using Kubernetes

A popular deployment target within the world of cloud applications is Kubernetes. Kubernetes acts as an orchestration layer for multiple components and/or microservices, providing mechanisms for resiliency and scalability. While this is a popular means of running microservice workloads, it is not always easy to set up locally to facilitate the development life cycle. Tools such as **minikube** offer a way to set up a lightweight Kubernetes cluster on your local machine, allowing you to test how that deployment might work. If you are running Docker Desktop, either on Windows or on macOS, there is an option to enable a Kubernetes Service to start when Docker starts. This can also give you a way to test out orchestration and deployments locally.

Having a target to deploy to is only half the battle, however. You still need to have a means of creating Kubernetes-compliant deployments that allow your code to run in that environment. You could read through the documentation for the platform and write deployment manifests from scratch. You could also use utilities such as **Helm** to create a bundled deployment using a standardized package of templates, but this will also require work on your part to ensure all of the services you need to have running are accounted for. Since we already have a compose file in the solution, it would be nice to be able to leverage that and not have to duplicate efforts. Fortunately, there are utilities that exist that can convert compose files into Kubernetes deployments.

One such utility is called `kompose`. This utility will take an existing compose file and transform that file into the appropriate YAML templates that Kubernetes needs to create your containers and Services in a cluster. Once you have installed a suitable local Kubernetes cluster, you will want to install `kompose` by referencing the installation instructions for your operating system at `https://kompose.io/installation/`. Running the utility is as simple as typing the following into a Command Prompt or console window:

```
kompose convert ./path/to/docker-compose.yml
```

This will create two files for each Service definition in your compose file—one for the Service itself and one for the Deployment. You can move all these new YAML files into a directory at the solution level called `kompose` to help track them in one place, or you could put the relevant Service and Deployment files into a `kompose` directory in each Service project directory. Take a moment to review the generated YAML files and compare them to the information in the compose file. You'll notice that items such as port mappings will show up in the service YAML, while others will appear in the deployment YAML.

After generating the YAML files, you'll want to test them out to ensure they are functional. If you're using local Kubernetes for the first time, the system will select that cluster as the current context. If you've connected to other clusters in the past, please make sure you set the context to the local instance (usually called docker-desktop) before testing the YAML files out. After the context is set, run the following commands to create a new namespace in your local cluster and apply the Deployment and Service files for the zookeeper and broker services:

```
kubectl create ns kafka
kubectl apply -f ./kompose/zookeeper-deployment.yml -n
  kafka
kubectl apply -f ./kompose/zookeeper-service.yml -n kafka
kubectl apply -f ./kompose/broker1-deployment.yml -n
```

```
    kafka
kubectl apply -f ./kompose/broker1-service.yml -n kafka
kubectl apply -f ./kompose/broker2-deployment.yml -n
    kafka
kubectl apply -f ./kompose/broker2-service.yml -n kafka
```

You can check the logs of the Zookeeper Pod via the command line (`kubectl logs -n equipment deployment/zookeeper`) to ensure Zookeeper is running and not experiencing issues. Feel free to experiment with the other Deployments and Services that were generated to get more comfortable with the makeup of the domain infrastructure.

Throughout this chapter, we've focused on setting up local environment support for the development of domain services, including how to enable container orchestration support. Next, we will look into an alternative way of managing development environments consistently through the use of GitHub Codespaces.

Leveraging GitHub Codespaces and Dev Containers

GitHub is the leading open source collaboration platform in the market today. Over 73 million developers use the platform, along with over 4 million organizations. Since being acquired by Microsoft in 2018, GitHub has continued to develop useful and innovative features to enable faster development cycles and remove barriers to collaborative work. One such innovation is GitHub Codespaces, which provides a free cloud-based development environment to any developer using any programming language. This can be a handy alternative to having to set up and script out a local development environment for a workstation or laptop. Let us check it out next.

GitHub Codespaces and online development

Given that Codespaces is a part of the GitHub platform, an obvious concern can be the fact that an internet connection will be required. Availability and bandwidth can vary by geographic area, and using an online developer environment may not be as beneficial as a local environment. Be that as it may, having the ability to run your environment in the cloud with no burden of maintenance on the environment itself can be very appealing.

Setting up a new Codespace is quite straightforward. By following the instructions found in the Codespaces Quickstart (see the *Further reading* section of this chapter for the link), you can be up and running with a new Codespace in a matter of minutes. GitHub will clone your repository into the environment and open an in-browser version of Visual Studio Code. Not only can you install extensions and have them persist between sessions, but you can also run and debug your application directly from the browser.

To test this out with the `Equipment` domain, create a new Codespace using the Quickstart link as a guide. Once the Codespace has been created and you see your editor in a new tab, navigate to the solution folder in the `src` directory of the repository by using the following terminal:

```
cd src/MTAEDA.Equipment
```

From there, choose one of the service projects within the solution folder, and change directories into it. For example, if you wanted to select the query service, you could type in the following:

```
cd MTAEDA.Equipment.Query.API
```

When you are in the directory of the service you wish to test out, simply type `dotnet run` in the terminal. This will kick off a restore, build, and ultimately run of the service. Once the service is up, you will see a popup window like the one in *Figure 5.3*. Clicking on **Open in Browser** will open a new tab with the base URL of that service, allowing you to navigate to the Swagger page (if relevant) or to perform another validation testing from the browser-based editor.

Figure 5.3 – Codespaces application popup

One drawback of using Codespaces is that there is only support for Visual Studio Code. This severely limits you if you are not using Visual Studio Code as your primary IDE. There is support for building and running these containers locally, though, which allows you to reap most of the benefits of Codespaces from the comfort of your desktop. This is where **Dev Containers** come in.

Dev Containers as a local environment

Using Dev Containers for your local development can be a convenient way to encapsulate everything you need into a centralized format. After creating a Codespace in the last section, a new folder will appear in the root of the repo named `.devcontainer`, which is where the configuration files for the Dev Container live. This directory can be copied and placed into the solution folder of any of the domains in the `src` directory. From there, the configuration files can be modified to tailor each container to the domain you'll be working with. You can use the default `devcontainer.json` file to define some language and IDE settings, or you can also include a Dockerfile to further customize your environment.

Now that we've explored a few different options for setting up the development environment, let's dive into Dockerfiles services a bit more.

Using Dockerfiles to build and run locally

Earlier in this chapter, we leveraged the Docker plugin for Visual Studio to generate Dockerfiles for each of the three services. As noted, the generated files contain several FROM directives, which allow Docker to use a specific image when building that layer. In some cases, Dockerfiles can be very simple—sequential lines installing or configuring aspects of the image for optimal use. But when does it make more sense to use a sequential file instead of one that has multiple build stages?

Sequential versus multi-stage files

Chances are, if you have Docker experience, you've constructed a few Dockerfiles in your day. Starting off, you may have taken a base image and added a few programs to it, installed specific frameworks on it, or even configured more complex options for the image itself. A good rule to follow when writing a Dockerfile is to look at what needs to happen during the build process itself. For example, when building images that will ultimately run a compiled program, .NET or otherwise, you will see multi-stage files in most cases. The reason for that is simple—it allows you to separate the building of the application from the packaging and configuration required to run the application.

Take the command API, for example. The Dockerfile generated has four stages in it, where it leverages two distinct base images. The intent is to ensure an optimal runtime image is used as the base and that an optimal SDK image is used to build the service. This gives each relevant stage a way to be separated but combined at the end of the process. When building an image for running a specific workload, you may opt to use a sequentially-ordered Dockerfile, as it may not need to keep things separated during the build phase. Take a moment and examine the three services' Dockerfiles that were generated earlier.

Now that we have a better understanding of how multi-stage Dockerfiles work, we can advance to the next step: wiring the services into our Docker Compose file to round out our environment.

Adding services to the Docker Compose file

For this example, we will use the Equipment domain services to stub out services in the compose file. At the end of the file, keeping indentations inline with the other services, begin by adding three new service definitions, as shown in the following code block:

```
query-service:
```

```
        image: ''
    query-service-db:
        image: ''
    command-service:
        image: ''
    event-listener:
        image: ''
```

For now, we will leave the image line as-is and move on to adding some details to each of the services. At a minimum, we will need to specify what port(s) are exposed for each service. This will help service discovery find different internal services and allow for attaching external tools to certain services. For example, exposing port 1433 on the query service database allows you to connect to the database using SQL Server Management Studio if you wish.

In some cases, services will have dependencies that must be running before the service can be brought up. Both the command service and the listener need to have at least one broker available, or the services will not function appropriately. The query service is only dependent on the database service. Using the depends_on attribute allows us to notate which services must be operational before **Docker Compose** can spin up the domain services.

There will also be a need to set up environment variables, which requires the environment attribute to be present in the service definition. For our example, only one of those services needs environment variables set—the database service. The SQL Server image needs to have a value for the **end-user license agreement** (**EULA**) set to **Y** for yes, or the container will not start up. In addition, a value must be supplied for the **system administrator** (**sa**) account, and that value is passed through an environment variable as well.

The hostname and container_name attributes are not required but do give you some flexibility in how the containers are named, both internally at the guest level as well as by Docker when creating them. We will fill those in with some easy-to-identify names to allow container identification either in the Docker UI or via the command line.

Having added these new fields to the service definitions, the following code block outlines how those attributes would line up under each service. Because the API services, as well as the listener, all run on a default port, we will map that default port to a unique port number for each service. This is also illustrated in the following code block:

```
    query-service:
        image: ''
```

```
      depends_on:
        - query-service-db
      ports:
        - 8081:80
      hostname: equip-cmd-svc
      container_name: equip-cmd-svc

    query-service-db:
      image: mcr.microsoft.com/mssql/server:2019-latest
      ports:
        - 1433:1433
      hostname: equip-event-db
      container_name: equip-event-db
      environment:
        ACCEPT_EULA: 'Y'
        SA_PASSWORD: $ADMIN_PASSWORD

    command-service:
      image: ''
      depends_on:
        - broker1
      ports:
        - 8082:80
      hostname: equip-cmd-svc
      container_name: equip-cmd-svc

    event-listener:
      image: ''
      depends_on:
        - broker1
      ports:
        - 8083:80
      hostname: equip-cmd-svc
      container_name: equip-cmd-svc
```

You'll also note that the variable for the sa password is listed as $ADMIN_PASSWORD. It will be important to set that to an appropriate value before bringing the containers up.

If you're using Linux or macOS, create a password that you can remember for the purposes of testing the database out and run the following:

```
export ADMIN_PASSWORD=<value of your password>
```

When you issue the command to start the containers via `Docker Compose`, this environment variable will be used during the build process to set the system administrator password for the database service.

Finally, let's point each of the images for the domain services to their respective directories using the `context` attribute. This will allow Docker Compose to build the images based on the path to the Dockerfile for each service. Change the `image` attribute to the `build` attribute and edit the line by removing the single quotes after the colon. Add a new line below it where the context attribute will be placed. As an example, the command API might look like the following:

```
build:
    context: ./MTAEDA.Equipment.API
```

By pointing at the directory for the command API, Docker Compose will automatically look for a Dockerfile in that directory to build the service's image. Repeat this process for the query service as well as the listener. Your image blocks should look like the following:

```
command-service:
    build:
        context: ./MTAEDA.Equipment.API

...

    query-service:
        build:
            context: ./MTAEDA.Equipment.Query.API

...

    event-listener:
        build:
            context: ./MTAEDA.Equipment.Listener
```

Test out the new configuration by running `docker compose build`, then `docker compose up`. You should see the services building where appropriate (our domain services) and pulling images where needed.

In this section, we have walked through the differences between single and multi-stage Dockerfiles and how to integrate the Dockerfiles and related services into your Docker Compose file for use locally.

Summary

In this chapter, we've looked at a few different ways to set up our development environment. We first looked at the local environment on our own computers, learning how to set up our infrastructure using Docker Compose as well as Kubernetes. Following that, we examined the use of online and portable environments using tooling provided by GitHub and Visual Studio Code. We explored how to generate Dockerfiles for our service projects and how to determine when a multi-stage Dockerfile makes sense. Using this new knowledge, you are equipped to build Dockerfiles to create images from your service projects as well as add targeted layers to the image build process to handle specific needs. Finally, we added our services to our Docker Compose file and started everything up in a single, cohesive stack.

In the next chapter, we will take that environment setup and put it through the paces by running, debugging, and testing those services using the environment we set up.

Questions

1. Why would you use Kubernetes over Docker Compose as an orchestrator for your infrastructure and your services?

2. What are some of the key fundamentals of containerization that can help improve the developer experience?

3. What is the significance of the `depends_on` directive in Docker Compose files?

4. Are there any limitations when using GitHub Codespaces as your development environment?

5. What's the difference between how the Kafka and database services are stood up versus the domain services?

Further reading

*   *Overview of Docker Compose* by Docker, available at `https://docs.docker.com/compose/`

*   *Kompose – Go From Docker Compose to Kubernetes* by Kompose, available at `https://kompose.io/`

*   *GitHub Codespaces QuickStart* by GitHub, available at `https://docs.github.com/en/codespaces/getting-started/quickstart`

*   *Codespaces Deep Dive* by GitHub, available at `https://docs.github.com/en/codespaces/getting-started/deep-dive`

- *VS Code Dev Containers* by GitHub, available at `https://github.com/microsoft/vscode-dev-containers`

- *Remote-Containers extension* by Microsoft, available at `https://marketplace.visualstudio.com/items?itemName=ms-vscode-remote.remote-containers`

- *Getting Started with Dev Containers* by Microsoft, available at `https://microsoft.github.io/code-with-engineering-playbook/developer-experience/devcontainers/`

6
Localized Testing and Debugging of Microservices

From the previous chapters, we understand that the **MTAEDA** application sample code currently contains a mix of runtime components: two self-coded microservices (**Producer** and **Consumer**), and three supporting services (**Zookeeper**, **Kafka**, and **Kafkadrop**). We saw that the Kafka service is configured with a level of redundancy. However, to leverage the full performance potential of **EDA**, we must be able to scale *all* the services by adding more instances in an orchestrated pattern.

This provides us with a configuration management challenge that is common across all modern application development practices today – not just for EDA:

- How do we develop and test efficiently with so many components to manage?

- Can we provide multiple developers and teams with a consistent environment?

- Will higher environments be reliably consistent with the development environment?

This chapter intends to walk through adding the configurations needed to *orchestrate* the development environment and using the tools that allow us to *test* code more effectively.

We'll cover the following topics in this chapter:

- Orchestration and containers
- Debugging in containers
- Testing against containers

By the end of this chapter, you will be able to orchestrate with containers all the components required for a complete local runtime environment, including those we are actively coding. You'll also learn how to attach to, and debug, code running in containers and know how to execute functionality and load tests against container-based microservices.

Technical requirements

For this chapter, you will require the GitHub source code found at `https://github.com/PacktPublishing/Implementing-Event-Driven-Microservices-Architecture-in-.NET-7/tree/main/chapter06`.

We will also be using **Postman**. If you have not used Postman before, but are familiar with making API requests using `curl` or `Invoke-WebRequest`, you should have no problem getting started with it. You can download the desktop version for free from `http://getpostman.com/`.

Another simple but effective open source tool we will use is **JMeter**, which the Apache Foundation manages and maintains. You can download it from `https://jmeter.apache.org/`.

Configuring orchestration and containers

We are already orchestrating components of the application using container technology. The `docker-compose.yml` file that we have explored in previous chapters defines the supporting containers we need and their configuration. When we execute `docker-compose` with this file, we are *orchestrating* the dependency resources by spinning up a specific configuration of containers in Docker.

The next step is to add the MTAEDA microservices to this orchestration definition, so we can develop code efficiently and consistently.

Everything as Code (EaC)

For a moment, let us validate *why* we want to continue with an orchestration approach to support the EDA development. I am confident you know at least the basics of **DevOps**! (I am referring to the practice, not the Microsoft service). By now, you may have also heard of, or worked with, **GitOps**. If you have not – do not fear. At the heart of both practices, you will simply find code to describe *every* element of a running workload.

That is exactly what we are going to do for this application development environment.

As developers, when we clone the application code, we want to ensure all configurations we need are in that same code base. As an operations team, we want to make sure we can deliver the environment that a developer expects for the application to function and perform successfully. The quality of the configuration code becomes as important as the actual application code itself.

If everything about the application, including the infrastructure requirements, is defined in code, we benefit in the following ways:

- Faster environment configuration
- Higher reliability
- Consistent testing
- Transparency of dependencies
- Easier portability

For this book, we will focus on the orchestration of the development environment so we can leverage the faster environment configuration. We must be able to efficiently start all application components while supporting continuous changes and debugging.

Creating container images

We will create a configuration file that describes how a container should host the application code for the Producer and Consumer services. To achieve this, we will create a Dockerfile for each service and use the Docker command-line interface to build and then launch these images.

Fortunately, **Visual Studio** can take care of most of this for us.

From the context menu on the consumer project, select **Add…** and then **Docker Support…**:

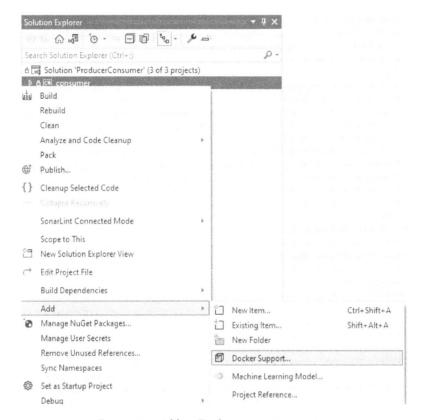

Figure 6.1 – Adding Docker support to a project

In the next popup, select **Linux** as **Target OS**.

Visual Studio makes simple but effective changes to the project. First, it adds a Dockerfile to the project. This file is automatically created with content to build the code in *Release* mode and then publish it to a final container image ready for launch. Let us explore that in a bit more detail:

```
FROM mcr.microsoft.com/dotnet/runtime:7.0 AS base
WORKDIR /app


FROM mcr.microsoft.com/dotnet/sdk:7.0 AS build
WORKDIR /src
COPY ["consumer/consumer.csproj", "consumer/"]
```

```
RUN dotnet restore "consumer/consumer.csproj"
COPY . .
WORKDIR "/src/consumer"
RUN dotnet build "consumer.csproj" -c Release -o /app/build

FROM build AS publish
RUN dotnet publish "consumer.csproj" -c Release -o
  /app/publish

FROM base AS final
WORKDIR /app
COPY --from=publish /app/publish .
ENTRYPOINT ["dotnet", "consumer.dll"]
```

The first FROM command points to a Microsoft container image called dotnet/runtime:7.0, which we alias with the name base. This image is a lightweight version of Linux with all the necessary installations to *run* pre-compiled **.NET 7.0** applications. It also sets the current working directory in that image to /app, which will be the destination for the compiled executable libraries later in the definition.

Before it can copy the application files to that destination, we must compile the code and publish it. To ensure we do not become dependent on any specific build requirements of our local machines, the build operation is executed in a container as part of the image creation process too.

The next section starts with a FROM command pointing to the dotnet/sdk:7.0 image with an alias of build. This is creating a new separate image based upon the Microsoft image that has all the necessary installations to *build* a .NET Core 6.0 application. It is worth noting that this image is proportionately larger than the runtime image, so not ideal as a base for the final release image.

The subsequent command lines copy source files, restore dependencies, and build and publish the code to the /app/publish directory against the build image.

The final section with the last FROM command takes the published (compiled) files from the build image and places them on the lighter base image. It then provides an ENTRYPOINT or *launch command* for when the container image runs.

The intermediatory build image is not persisted and, therefore, does not add overhead to the final container image created.

Beyond creating the Dockerfile for us, Visual Studio has also added a package reference to the project file and property to persist our **Target OS** choice. You can see these changes in the .csproj file.

Finally, Visual Studio added a new launch profile in the launchSettings.json file under the Properties folder, simply named Docker.

We are now ready to build the Consumer application as a container image and run it. This is not something we will do manually in the future, but at this stage, it will help with the understanding of creating container images for reuse.

From a new command prompt in the solution folder, run the following:

```
docker build . -f consumer\Dockerfile -t consumer:latest
docker run -d consumer:latest
```

Running these commands, you will be able to follow the image build process as described previously, and finish with a running container based on that image.

Open the Visual Studio container monitoring window by clicking on the **View… | Other Windows… | Containers** menu items. From here, you can see the running container and inspect its output logs.

Do not worry if the container logs show errors at this point. We have not yet created a Producer container or orchestrated it with the other services in the existing docker-compose.yml file.

At this point, you should repeat all the steps in this section for the Producer project. In each step, simply replace any references to the consumer with the producer. Once completed, your solution should have two separate Dockerfiles, each in their respective project folder, and a successful build for each, as shown here:

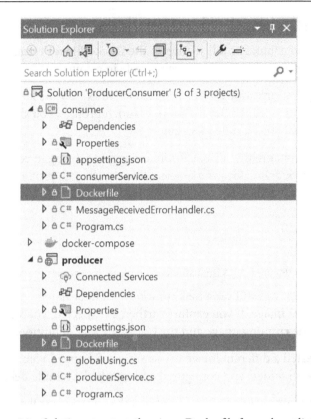

Figure 6.2 – Solution structure showing a Dockerfile for each application

Now that we have the container images, we need to focus on using these to debug our code inside them. When we orchestrate many containers, including the service containers, we will then retain the ability to break into, and step through, lines of code in a production-like environment.

Debugging in containers

The container images can be started on demand to run the microservices. However, producing a new Docker image that internally builds and publishes *release* code will not offer the debugging of the code we are looking for!

To overcome this, we need to run the development code unpublished, using a **Debug** build, quickly and dynamically inside a container with a debugger attached. We must understand how Visual Studio modifies the container images to support debugging, and how this fits into the overall orchestration of all components, including the service component.

Debugging individual microservices

While Visual Studio did a fantastic job setting up the Dockerfile for release builds, it does not use this *entire* file definition when debugging code in a container.

The additional configurations added to the project enable exactly the kind of *fast and dynamic* environment we need. Under the project launch profiles, you should now see a new one added by Visual Studio called **Docker**.

Before we look to understand how Visual Studio manages to debug in containers, we first want to confirm the debugging does work. Select this new launch profile and press the **Debug** play button (or hit *F5*):

Figure 6.3 – Visual Studio debug launch controls

In the container window, you will see a new running container named `consumer` based on the `consumer:dev` image. If you explore further, as expected, you will see noticeable differences between the running image and the `Dockerfile` definition.

Visual Studio has created a different image with a series of mapped volumes (including one to the local solution folder) and configured and attached a remote debugger.

You can stop the debugging session at any point and the container will continue to run, even though the application process has stopped. You can add breakpoints and relaunch the debugging session as you would in any local debug session, knowing that the processes are instead running within a container.

We want to control the development environment dependencies as code via `Dockerfile`, but we also just discovered that Visual Studio does not use that Dockerfile when launching the debugger. So, how do we customize that environment?

Visual Studio offers flexibility to do this. If we want to add a customization that affects both the final release image and the debugging environment, we add commands for the base section of `Dockerfile`:

```
FROM mcr.microsoft.com/dotnet/aspnet:7.0 AS base
WORKDIR /app
EXPOSE 80
EXPOSE 443
# <add commands here to modify both the debug and release
   container image>
RUN dotnet tool install --global dotnet-ef
```

When we run a debug session, Visual Studio will make sure the custom container includes everything from the base image definition.

If we want to add commands that only apply to the debugging image, but not the final release, we need to add another image – created from base:

```
FROM mcr.microsoft.com/dotnet/aspnet:7.0 AS base
WORKDIR /app
EXPOSE 80
EXPOSE 443
# <add commands here to modify both the debug and release
   container image>
RUN echo "This is my base image" > /tmp/image_type


FROM base as mydebug
RUN echo "This is my debug image" > /tmp/image_type


<section omitted for brevity>
FROM base AS final
WORKDIR /app
COPY --from=publish /app/publish .
ENTRYPOINT ["dotnet", "consumer.dll"]
```

You can see that the final image is based on the base image, so it will not include any of the commands in the mydebug image. In the example commands added, we expect our base image (or any image deriving from it) to have a file called /tmp/image_type with the content, This is my base image. When running our debug container, we expect Visual Studio to have targeted the mydebug image, which should instead have the content, This is my debug image, in the /tmp/image_type file.

The final thing we must do is tell Visual Studio the name of this new image to use in place of base when debugging. To do this, add the following section to the .csproj file:

```
<PropertyGroup>
      <!-- other property settings -->
      <DockerfileFastModeStage> mydebug
          </DockerfileFastModeStage>
   </PropertyGroup>
```

Hurrah! Managing dependencies for the development environment, and any specifics for a debug environment, are now under source control!

We can be sure that all developers using this same code base and the Docker launch profile are running a consistent container definition matched to the development and production needs. So, we can launch a *new* fully working debug session on *any machine* that has Docker and Visual Studio installed, without needing any other local dependencies. This even extends to the latest .NET Core SDK version you may or may not have installed. If you ever like to play around with preview releases of the SDK, you can appreciate how useful this is.

The portability of the MTAEDA development environment is much simpler, and now reliably controlled via code. Remember the *Everything as Code (EaC)* section?

But wait…we are only debugging one microservice at a time. And what about the Kafka cluster services?

Orchestrating and debugging all services

We have the microservices individually running, and debugging, in containers. In order to debug them in a useful way, we must also have the Kafka services running for the application code to fully work. We configured Zookeeper, Kafka, and Kafdrop in the existing `docker-compose.yml` file. So, how do we bring this together with the newly created microservice Dockerfiles?

We want Visual Studio to set up a new `docker-compose` project. In doing so, it will try to create a new `docker-compose.yml` definition file. To avoid conflict, we must back up the existing file with the Kafka services before doing so.

Simply rename `docker-compose.yml` to `docker-compose.bak`.

Next, select **Add** and then **Container Orchestrator Support…** from the context menu on the `consumer` project:

Figure 6.4 – Adding container orchestrator support

Chose the option for **Docker compose** and **Linux** in **Target OS**.

You will see a new project type added to the solution called `docker-compose`, and within that, we have a new `docker-compose.yml` file:

```
services:
  consumer:
    image: ${DOCKER_REGISTRY-}consumer
    build:
      context: .
      dockerfile: consumer/Dockerfile
```

In that file, you will see a single service added that points to the `consumer` Dockerfile as the image to build.

Check the **Containers** window again and you will see a new container with a remote debugger configured and the application code ready for a debug session:

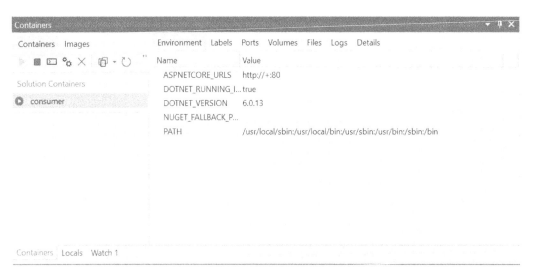

Figure 6.5 – Containers window showing consumer running as defined in the docker-compose.yml file

The only real difference now is that Visual Studio understands that we want *everything* in the new `docker-compose.yml` file up and running, rather than one application at a time, as we had with the Docker launch profile.

So, now we can add the Producer service to the file in the same way. We will also add a port mapping to expose the API to the local machine. By default, the Producer exposes an API endpoint on port `80` on the container, and we will map that to port `5000` on the localhost:

```
services:
  consumer:
    image: ${DOCKER_REGISTRY-}consumer
    build:
      context: .
      dockerfile: consumer/Dockerfile

  producer:
    image: ${DOCKER_REGISTRY-}producer
    build:
      context: .
```

```
      dockerfile: producer/Dockerfile
    ports:
    - «5000:80»
```

Now, we can add back the Kafka services, copying them from the docker-compose.
bak file:

```
services:
  consumer:
    image: ${DOCKER_REGISTRY-}consumer
    build:
      context: .
      dockerfile: consumer/Dockerfile

  producer:
    image: ${DOCKER_REGISTRY-}producer
    build:
      context: .
      dockerfile: producer/Dockerfile
    ports:
    - «5000:80»

  zookeeper:
    image: bitnami/zookeeper
    hostname: zookeeper
      ...        <omitted for brevity>

  broker1:
    ...        <omitted for brevity>

  broker2:
    ...        <omitted for brevity>

  kafka-ui:
    ...        <omitted for brevity>
```

The final changes we must make are to the appSettings.json files. In previous chapters, all the services were running locally, so we used localhost to address the Kafka brokers. Now, as each service runs in its own container, we must address them by their container hostname within Docker and their *internally* available ports:

```
{
  «Topic»: «equipment»,
  «KafkaServer": "broker1:9092",
  «DefaultGroupId": "consumers.equipment"
}


{
  «Topic»: «equipment»,
  «ProducerConfig": {
    «BootstrapServers": "broker1:9092"
  }
}
```

Now, we can select the new docker-compose project and press the **Debug** play button (or hit *F5*):

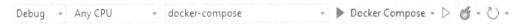

Figure 6.6 – Visual Studio debug launch controls

To prove everything is up and running as expected, we can send a message to the Producer API endpoint with this command:

```
Invoke-RestMethod -Method Post -Uri "http://localhost:5000/
send" -UseBasicParsing -Body "This is a test to our
orchestrated environment"
```

You can observe the Producer log processing the POST request, and the Consumer log echoing the message it picked up from the event stream. Of course, we can also set breakpoints at any point in either of the applications.

If we look again at the **Containers** window, we can see all the containers up and running as expected:

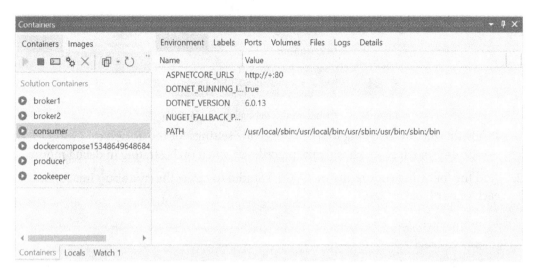

Figure 6.7 – Containers window showing all services running as defined in the docker-compose.yml file

We now have a fully debugging development environment, running parallel microservice code, supported by Kafka services.

When we stop this debug session, Producer and Consumer will both terminate, but all containers will remain running.

It is also worth noting that Visual Studio will attempt to shut down and delete running containers if you close the solution or exit the IDE.

Fixing the Debug custom image

When progressing from single container debugging to orchestrator debugging, we lost a very important feature from our previous section, *Debugging individual microservices*. The containers defined in the docker-compose file are no longer applying the custom debug build commands we added in the Dockerfile image definitions that we named mydebug.

We modified the .csproj file and added a setting called DockerfileFastModeStage to specify the name of the image that should be built and used *only* when running in debug mode. Unfortunately, this is not honored by Visual Studio when using docker-compose to orchestrate multiple containers.

Fortunately, the fix is straightforward:

1. In the Visual Studio docker-compose project, add a new file named docker-compose.vs.debug.yml. There may be an existing file named very similarly to this, but please ensure you have a new file with this specific filename:

Figure 6.8 – Visual Studio debug launch controls

Inside this file, we are going to specify override settings for the `docker-compose.yml` file, which will take precedence when orchestrating in debug mode.

2. Add the following code to instruct Visual Studio to target the `mydebug` images in each of our Dockerfiles:

```
version: '3.4'
services:
  consumer:
    build:
      target: mydebug

  producer:
    build:
      target: mydebug
```

Debugging the `docker-compose` project will now target the correct image definition in the Dockerfiles for debug mode.

Now that we can quickly spin up the entire development environment on any machine with Visual Studio and Docker installed and start debugging, we can turn our attention to testing!

Testing against containers

A full testing regime across all stages of a DevOps life cycle is not in scope for this book. However, we will walk through the technicalities of performing basic testing to prove the environment is healthy and demonstrate putting a load on the overall application. We want to show the high-scale throughput we can achieve by adopting an event-driven architecture over others.

Functionality testing

We will set up a simple test using Postman:

1. Create a new collection, which is simply a folder to store a group of related API requests that we can execute repeatedly.

2. Next, create a new request. Set the method type as POST and set the request URL as `http://localhost:5000/send`.

3. Select the **Body** tab, change the format to **Raw**, and enter an event message in the text area. I chose to use This is a Postman test to our orchestrated environment:

Figure 6.9 – Setting up a simple Postman API request

4. Make sure the Visual Studio solution is open and you are debugging using the `docker-compose` launch profile from the last section.

5. Click **Send** and you should see Postman confirm that it received a Status code is 200 response. If you inspect the container logs for Producer and Consumer, you will see the message proliferated.

    Postman has a powerful set of JavaScript-based scripting tools, allowing you to add scripts for pre-request operations and post-request test operations.

6. In the **Tests** script box, add the following code, which allows you to assert that the response was successful:

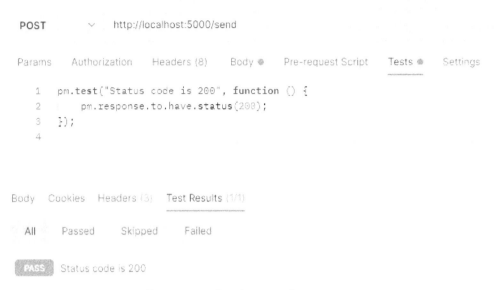

Figure 6.10 – Creating a simple response test

You can also process any response body and use this to populate environment variables for the next test you may create. These are all excellent Postman features that I encourage you to explore, as their patterns translate to most other API testing services and solutions.

We can take this opportunity to confirm the fault tolerance of the design. If you stop the `broker1` container, you will see that the consumer fails to monitor events.

To stop the container, select it in the **Containers** window and click the **Stop** icon:

Figure 6.11 – Stopping the broker1 container via the Containers window

We can keep executing the Postman test, and the messages will continue to be produced successfully on the Kafka cluster as events. Restarting `broker1` allows the consumer to catch up on missed events and process them as normal.

Postman offers a cloud service to automate the testing of published APIs. It's worth noting that you can also export Postman collections as `.json` files and use an npm package called `newman` to execute the tests from a command line (`https://www.npmjs.com/package/newman`). Adding these tests as `.json` files to a repository is a terrific way to drive routine testing into a build pipeline. There is an example of this format included with the source code. Explore the `PostmanExportExample.json` file in the root of the `chapter06` folder.

Load testing

Postman testing is great to confirm whether API endpoints are behaving as expected; however, we want to put a heavy load on the Producer to ensure the design of the application will keep up and process heavy throughput traffic. There are many tools we can use to do this. At a more mature point in the development process, we may decide to use a service such as **Azure Load Testing** (in preview at the time of writing), which can bombard the application at scale and produce millions of metric points to help us understand performance and any unintended bottlenecks caused by anti-patterns.

To demonstrate the technicalities of implementing load testing, we will use **JMeter**.

The steps to set up a request in JMeter are similar to Postman, except we have a few additional configurations to add the scale of requests:

1. Under the default test plan, add a thread group:

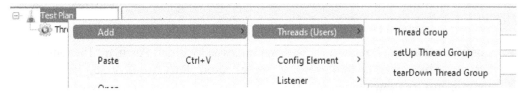

Figure 6.12 – Adding a thread group to a test plan

2. Configure this thread group to simulate 500 simultaneous threads (users) and leave all other settings as default:

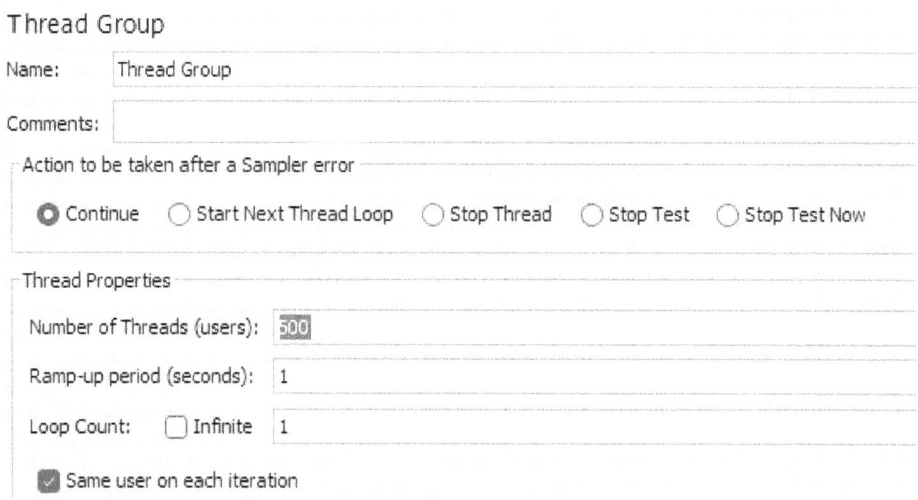

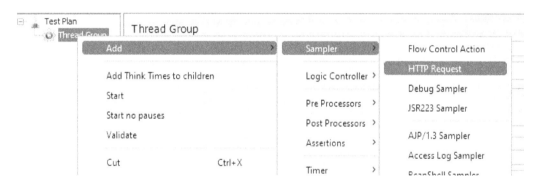

Figure 6.13 – Configuring a thread group

3. Beneath **Thread Group**, add a new **Sampler | HTTP Request** option:

Figure 6.14 – Adding an HTTP request sampler to a thread group

4. Configure this HTTP request to the same values as we did in Postman. The request URL must be broken up into **Protocol**, **Server Name**, **Port Number**, and **Path** components, as shown here:

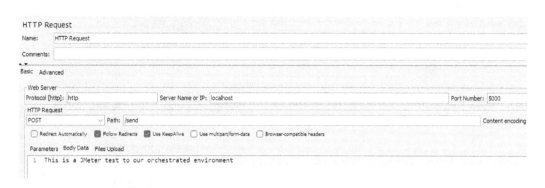

Figure 6.15 – Configuring an HTTP request

5. Beneath the HTTP request, we want to add a summary report that will show us the performance of the load test when running:

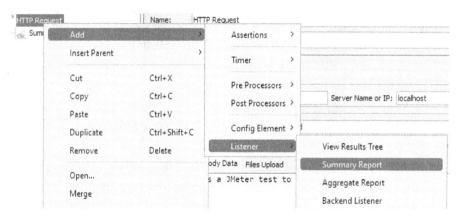

Figure 6.16 – Adding a summary report to an HTTP request

There is nothing to configure on this item.

6. Now, with **Summary Report** selected, we can click on the green **Start** icon on the main toolbar and watch 500 simultaneous threads put a load on the Producer's /send endpoint:

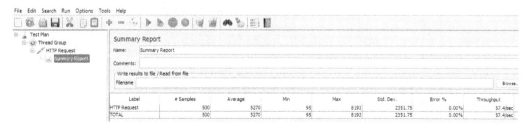

Figure 6.17 – Summary Report output

In my summary report, I achieved a throughput rate of **57.4** requests per second. This does not sound like much, but be aware that we have not scaled up the Producer application yet.

We expect the Kafka cluster to take higher throughputs when recording new events. So, when we scale out the Producer to more instances, we will increase that overall request throughput.

The Consumer will also scale and, with that, we can demonstrate an important level of overall resiliency. We can take down the Consumer application and still load test the Producer, with certainty that the events are recorded and ready for the Consumer to read whenever it returns to a healthy state.

Summary

In this chapter, we have detoured away from the application itself to focus on setting up an efficient and reliable development environment – one that supports running multiple support services while also debugging multiple code projects in the solution. We touched upon how this fits well with modern full life cycle application development using DevOps and GitOps practices.

We also looked at how we can set up functional API tests using Postman and heavy load tests using JMeter. With a fully debuggable environment, fixing broken functionality and improving code performance becomes easier.

In the next chapter, we will start to look at the observability of the microservices in the application. This includes application health, performance, and availability.

Questions

1. What pipeline-like activity does Visual Studio provide in its default Docker template for a project?
2. How does Visual Studio efficiently debug code in a container?
3. What would happen if the Producer service failed entirely but all other components were healthy?

Further reading

- *GitOps and the Rise of Everything-as-Code*, by Jelle den Burger, available at `https://www.meshcloud.io/2022/04/04/gitops-and-the-rise-of-everything-as-code/`

- *Visual Studio Container Tools*, by Microsoft, available at `https://docs.microsoft.com/en-us/visualstudio/containers/overview?view=vs-2022`

- *Modify Container Images Only for Debugging*, by Microsoft, available at `https://docs.microsoft.com/en-us/visualstudio/containers/container-build?WT.mc_id=visualstudio_containers_aka_containerfastmode&view=vs-2022#modify-container-image-only-for-debugging`

- *Create a multi-container app with Docker Compose*, by Microsoft, available at `https://docs.microsoft.com/en-us/visualstudio/containers/tutorial-multicontainer?view=vs-2022`

- *Docker Compose build properties*, by Microsoft, available at `https://docs.microsoft.com/en-us/visualstudio/containers/docker-compose-properties?view=vs-2022`

- *API Testing Strategy*, by Cerosh Jacob, available at `https://medium.com/geekculture/api-testing-strategy-384af1ff43f6`

- *Event-Oriented Architecture Anti-Patterns*, by Sergi Gonzalez, available at `https://medium.com/letgo/event-oriented-architecture-anti-patterns-2dccc68ed282`

7
Microservice Observability

In the last chapter, we were able to *orchestrate* the **MTAEDA** application so that it is easy to launch and debug all the components of the system. Every individual service is available for interrogation at runtime, so it is easy to identify issues and gain the necessary insight to resolve them.

Eventually, the MTAEDA application will reside in several other environments outside of active development. When the application doesn't work as expected, it is vital to be able to observe what is going on internally. Classically, monolithic applications produce cohesive serial logs that can be observed as needed.

Observability is critical to the speed of the overall development life cycle in **quality assurance** and **user acceptance testing** environments. When more effort is required to find out what is happening during an application failure, this ultimately prolongs the development of a fix and the subsequent testing to verify that fix. DevOps cycles are ground to a halt when whatever is happening in relation to a bug can't be observed quickly. In production environments, often, the only way a bug may *ever* be observed is through the series events of happening *in the wild*. These difficult-to-reproduce production bugs are critical to resolving an application and it being viewed as well supported and reliable.

This chapter intends to explain how to shift the mindset of observability relative to monolithic software development – first, by addressing the challenges through using **microservice** patterns, then by addressing the challenges introduced by **event-driven** architecture.

We'll cover the following topics in this chapter:

- How to monitor service health and availability

- Correlating logs across microservices

- Using causation logging across events

- Leveraging Azure Application Insights for observability

By the end of this chapter, you will know how to add independent logging to event-driven application services *without losing* the aggregated observability needed to support higher environments.

Technical requirements

For this chapter, you will require the GitHub source code found at `https://github.com/PacktPublishing/Implementing-Event-Driven-Microservices-Architecture-in-.NET-7/tree/main/chapter07`.

We will also create an instance of Application Insights in Azure. Details about this service, which is part of the Azure Monitor service, can be found at `https://azure.microsoft.com/en-us/services/monitor/`.

Observability

It would be prudent to start with a clear definition of observability. In other words, it means *to be able to notice or discern something*. Applied to our focus on software applications, we can be more specific about what we are discern and how we do so: *to be able to measure the internal state of a system by its outputs*.

In software systems, this is achieved through the enablement of what is commonly referred to as the **three pillars of observability**:

1. **Metrics**: A series of measurements over time

2. **Logs**: A record of messages describing noticeable events within a system

3. **Traces**: A set of indicators throughout logs that connect a series of related events

At this point, it is worth addressing the critical commentary you will find on the three pillars of observability in more recent online publications. There *are indeed* some shortcomings of these pillars at a hyper-scale, and thought leaders are evolving beyond them to refine the quality and accuracy of observability even further. However, understanding and using the three pillars is a fundamental requirement to be able to go beyond them. They are entirely relevant and valuable to the MTAEDA application and implementing the basic principles is required for all microservice architectures.

In the world of EDA, we have two significant complications to overcome that require a modified approach to implementing these pillars:

1.   Applications with a microservice architecture are, by definition, made up of many smaller separate components that run in total process isolation from each other. They may even run on different infrastructures, be developed with different languages, or require different CPU architectures.

2.   Event-driven architecture decouples microservices to the point that we cannot control how many downstream consumers react to a single produced event.

For each of the three pillars, let us understand the specific challenges introduced by these complications, as compared with monolithic applications. This will provide the key questions to answer in subsequent sections of this chapter.

Metrics

An application can produce a series of **metrics** that make it more observable. The primary and most critical metric is whether the application is running. For classic monoliths, this is generally not a difficult thing to observe. If the application is not running or has internally stalled, it will not process any more work. A monolith may be monitored by an external service to check that it is still running, often referred to as a health probe or availability signal.

In the world of microservices, we cannot (and should not) think of a single availability signal for the entire application. The application is made up of many independently running services, each with its own ability to be healthy – or not. One service being unavailable does not mean a total failure of the overall system. Other services should be **fault-tolerant** and still function. It may be that continuing to function is restricted to responding with a *dependency failure* message, but the point is that is it still responsive and able to function in some way.

Furthermore, microservices should scale horizontally. While one instance of a service may be unhealthy or unavailable, *it should not* represent a single point of failure for the overall system.

Other metrics become important when we start to look at the performance, or throughput, of a service. Where a monolith can report CPU usage as a binding limit – 100% and your application is maxed out – a microservice landscape is monitored as a mix of CPU values, representative of the demand placed on specific smaller chunks of the overall application functionality and providing a useful method to drive horizontal scaling. In addition to the CPU, an instance of a single service may emit its memory usage, storage performance, or even a custom metric that is relevant to its purpose (for example, how many concurrent widgets it is processing).

For this chapter, we will focus on the availability or health metrics of a service. More specifically, these are referred to as **liveness** and **readiness** probes. We will explore these more in a later section.

Logs

An application outputs a level of **verbosity** about its internal state at specific events. This is not to be confused with the events in event-driven architecture. This kind of event can be any line of software code that is executed that emits a message. There is typically a setting exposed to dictate the application's level of verbosity, typically ranging from very little output (**informational**) to a continuous stream of messages (**trace**). As we move from development to higher environments, the verbosity setting tends to reduce the output unless there is an explicit need to observe more detail for an elusive bug!

In the world of microservices, the challenge for logging is one of aggregation and standardization. How do we bring all these distributed logs together so we can observe the system as a whole? How do we avoid the overhead of every service instance hosting its own logging system? How do we aggregate at a service level while still having granularity for a specific service instance?

Later, we will answer these questions and cover the observability of logs.

Traces

The use of the word *trace* here should not be confused with the verbosity level of logs. In a single large application, viewing the most verbose log information may indeed deliver a trace of events that can be tied together. However, even in the most archaic single-threaded monolith, you will often see correlation identifiers in the log messages. This provides the *real* definition of a trace. For any end-to-end system operation (input to output), there should be a unique identifier that links together any message output from all the functions that execute. This is the correlation identifier.

In the world of microservices, traces follow the same approach of using a unique identifier to tie together different executing services and functions within them, so we can trace all events related to a single originating system action.

While microservices themselves are expected to be small and relatively simple in implementation, the overall system can be extremely large and highly complex, to the point that a single correlation identifier can return an enormous event log trace. Querying logs by this value may return more data than we can reasonably understand, making it very difficult to quickly identify an internal problem. While it is possible to reduce that footprint by querying for a specific service within the logs, this is of no use when we are uncertain of which service the problem may arise in.

When we embrace event-driven architecture, the correlation scale issue becomes even greater. A system may write many logs for a single correlation identifier spanning many services and then produce an event in the event stream. That event stream may be subscribed to by any number of consumers. This is unknown to the producer. However, the downstream consumers could be either no applications or thousands of applications! Assuming there are many, they would propagate the same correlation identifier, in which case a single trace may explode exponentially. We will cover how to interrogate traces within logs better during this chapter.

Now that we have a good understanding of the three pillars of observability and how they are complicated by highly scalable systems, such as those built with event-driven architectures, we can move on to implementing metrics for service liveness and readiness.

Liveness and readiness

Every component service in our application should provide liveness and readiness metrics as a minimum. In short, a liveness metric advertises that a service is functioning, and a readiness metric advertises that a service is available for requests. When a service is managed by an orchestrator, specifically Kubernetes, its definition can include references to endpoints that individually report these metrics. The endpoints do not need to be exposed outside of a pod.

If a **liveness** endpoint indicates the failure of the service or doesn't report anything at all, the orchestrator will *terminate the pod* and replace it with a new instance.

If a **readiness** endpoint indicates the container is unavailable, the orchestrator will *stop sending request traffic* to it.

Note that these endpoints *will not* have any useful functionality in our development environment with `docker-compose`. **Docker** is a simplified orchestrator that focuses on running multiple containers on a single machine. **Kubernetes** is a broader functioning orchestration platform that includes more complex frameworks, such as traffic control and monitoring service health over multiple compute nodes.

Creating liveness endpoints is relatively simple.

Liveness endpoints

It is possible to configure a liveness probe in Kubernetes using any three different methods:

1. A command probe
2. An HTTP request probe
3. A TCP probe

For the **Producer** service, we will focus on creating a Liveness API endpoint, which will be configured with an **HTTP request probe**.

For the **Consumer** service, we will focus on creating a TCP listener service, which will be configured with a **TCP probe**.

Adding the Producer liveness endpoint

In the `Program.cs` file of the producer project, there are already inline definitions for simple endpoints using the `MapGet()` and `MapPost()` methods. Add the following between the `"/"` and `"/send"` setup code:

```
app.MapGet("/healthz", () => "Ok!");
```

It quite simply creates a new API endpoint that returns a string of `Ok!`.

Launch the application and access the health endpoint via `http://localhost:5000/healthz`:

Figure 7.1 – Response from the liveness health API endpoint

The default response code for `MapGet()` is `200`. Regardless of what content is returned, Kubernetes will interpret that response code as the service being alive and well. If our application hangs or there is an API service problem, we expect this endpoint to fail with an exception, returning a response code of **500**, or not to respond at all. Both scenarios will inform Kubernetes that the service instance is unhealthy and must be terminated and replaced.

What if the logic behind the liveness of the service instance is more complicated? We will add another endpoint to demonstrate a more complex logic to determine the liveness status and programmatically set the response code:

```
app.MapGet("/complexhealthz", (HttpContext http) =>
{
    if (DateTime.Today.DayOfWeek == DayOfWeek.Sunday)
        http.Response.StatusCode = 500;
    else
        http.Response.StatusCode = 200;
    return $"It's complicated! {http.Response.StatusCode}";
});
```

Relaunch the application and access the liveness endpoint via `http://localhost:5000/complexhealthz`.

If the current day of the week is Sunday, it will respond with a status of **500**, indicating it is unhealthy:

Figure 7.2 – Response from the complex liveness API endpoint indicating an unhealthy state

Otherwise, it will respond with a status of **200**:

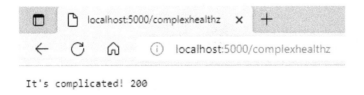

Figure 7.3 – Response from the complex liveness API endpoint indicating a healthy state

Let us examine what the portion of the Kubernetes manifest configuration looks like to use the liveness HTTP request endpoint:

```
spec:
  containers:
  - name: procuder
    image: mtaeda/producer:1.0
    livenessProbe:
      httpGet:
        path: /healthz # or /complexhealthz
        port: 80
      initialDelaySeconds: 3
      periodSeconds: 5
      failureThreshold: 2
```

This is straightforward to interpret. It tells Kubernetes the path and port to look for the liveness endpoint. The `initialDelaySeconds` parameter is a grace period to allow the service to start up, so it isn't immediately seen as unhealthy and terminated if there is any delay before the API endpoint is launched. The `periodSeconds` parameter specifies how frequently the liveness endpoint is checked. The `failureThreshold` parameter specifies how many times the liveness endpoint must sequentially fail to trigger termination and the replacement of the pod.

Adding the Consumer liveness endpoint

Adding a TCP liveness endpoint is slightly more involved for the consumer. In the `Main` entry point, add the following code before building `Ihost`:

```
Task.Run(() => {
    var tcpHealthEndpoint = new TcpListener(IPAddress.Any,
      123);
    tcpHealthEndpoint.Start();
    while (true)
    {
        TcpClient client = tcpHealthEndpoint
          .AcceptTcpClient();
        NetworkStream stream = client.GetStream();
        var msg = System.Text.Encoding.ASCII.GetBytes("Ok!");
        stream.Write(msg, 0, msg.Length);
        client.Close();
```

```
    }
  });
```

This code creates a TCP listener on port `123` bound to all network interfaces attached to the system that the process is running on. The `AcceptTcpClient()` method is a blocking call, which is why the entire scope is wrapped in a `Task.Run()` method. Otherwise, it would block the main consumer service from starting. The example writes out `Ok!` when a TCP client successfully connects. This is an optional step. All Kubernetes requires to confirm liveness is a successful connection, even if the connection is immediately closed.

This change is sufficient for a Kubernetes deployment, as the orchestrator can access the container pod directly and interrogate port `123`. To test this locally with `docker-compose`, we must project that same port to our localhost. Modify the `docker-compose.yml` file to include a port mapping from the consumer:

```
services:
  consumer:
    image: ${DOCKER_REGISTRY-}consumer
    build:
      context: .
      dockerfile: consumer/Dockerfile
    ## Add this port mapping
    ports:
    - "123:123"
```

Restarting the `docker-compose` debug will apply this change. Now, start a command prompt and attempt to connect via `telnet` to the port:

```
> telnet localhost 123
```

We get the following output:

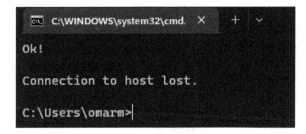

Figure 7.4 – Telnet output from a healthy liveness TCP endpoint

Let us preview what the portion of the configuration looks like to instruct Kubernetes to use our liveness TCP request endpoint:

```
spec:
  containers:
  - name: consumer
    image: mtaeda/consumer:1.0
    livenessProbe:
      tcpSocket:
        port: 123
      initialDelaySeconds: 15
      periodSeconds: 10
      failureThreshold: 2
```

Just like the configuration for an HTTP liveness probe, we can configure an initial delay, how frequently we check, and the threshold before the orchestrator takes action on an unhealthy service instance.

Implementing complex liveness logic using `TcpListener` is more involved, requiring listeners to be dynamically started and stopped, and that is beyond the scope of this book. However, there is a link in the *Further reading* section to an example of how this can be achieved.

Now that we have successfully added liveness endpoints, we should move on to understanding readiness endpoints and implementing them too.

Readiness endpoints

Just like the liveness endpoints, Kubernetes can probe readiness endpoints using an HTTP request, TCP connection, or direct commands. The successful implementation of a readiness endpoint will make Kubernetes stop sending traffic to an instance of our service if it advertises that it is healthy, but not ready to accept requests.

Adding the Producer readiness endpoint

Let us simulate a scenario where the Producer service starts, but we don't want to accept any requests for the first 30 seconds. In the real world, some services may have a lot of work to do on startup, and even though the liveness endpoint is responsive, we may want to hold off incoming requests until another operation completes.

Beneath the /complexhealthz endpoint, add the following code:

```
var appStartTime = DateTime.UtcNow;
app.MapGet("/readyz", (HttpContext http) =>
{
    var readyAt = appStartTime.AddSeconds(30);
    if (DateTime.UtcNow < readyAt)
    {
        http.Response.StatusCode = 500;
        return $"Ready in {(readyAt -
          DateTime.UtcNow).TotalSeconds} seconds.
             {http.Response.StatusCode}";
    }
    else
    {
        http.Response.StatusCode = 200;
        return $"Ready! {http.Response.StatusCode}";
    }
});
```

Launch the application and access the readiness endpoint via http://localhost:5000/readyz.

Within the first 30 seconds of starting, the response code will be **500** and the message will display how many seconds are left until the service is ready to accept requests:

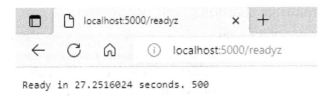

Figure 7.5 – Readiness HTTP endpoint showing not ready to service requests

After 30 seconds, the response code will be **200** and the message will confirm the service is ready to accept requests:

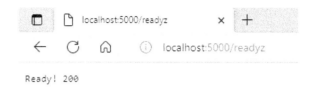

Figure 7.6 – Readiness HTTP endpoint showing it is ready to service requests

Let us preview what the portion of the configuration looks like to instruct Kubernetes to use our liveness TCP request endpoint:

```
spec:
  containers:
  - name: consumer
    image: mtaeda/consumer:1.0
    readinessProbe:
      httpGet:
        path: /readyz
        port: 80
      initialDelaySeconds: 3
      periodSeconds: 5
      failureThreshold: 2
```

Remember that Kubernetes does not care about the message content that is returned, only a status code of **200**, meaning ready, or any other code, meaning not ready.

Skipping the Consumer readiness endpoint

It does not make sense to add a readiness endpoint to the consumer service. It does not accept incoming request traffic that could be blocked if the service is deemed not ready. Instead, it monitors the event queue itself for events to fire against. Therefore, its readiness state, and control, are both internally self-determined.

Now that the liveness and readiness endpoint metrics are in place, we can start to focus on aggregating logs.

Aggregation of logs

When using Kubernetes, we can think of two primary methods for emitting logs. The first is directly from our code to an external logging system. The second is direct output via the standard output and error streams.

To aggregate logs directly to an external logging system, we can use **Azure Application Insights**. It can natively instrument **.NET** applications (among many other supported languages) with a few code changes.

To aggregate logs written to the standard output and error streams, we can use **Container Insights**, which is a capability within **Azure Monitor**. Container Insights is a great utility for viewing and querying logs produced by multiple Kubernetes clusters, but it currently does not easily support the level of structured interrelationship logging that this application will generate.

The MTAEDA application will use telemetry and logging directly to Application Insights so that we can support tracing.

Now, let us set up an Applications Insight instance to manage the log aggregation.

Creating an Azure Application Insights instance

Head to https://portal.azure.com and sign in with your user account to access a usable subscription, as described in the *Technical requirements* section.

Type and find Application Insights in the main search box:

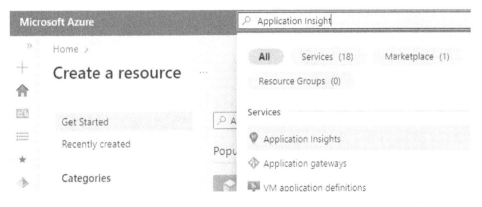

Figure 7.7 – Finding Application Insights instances in Azure

Unless you have used Application Insights before, the list of instances should be empty. Click on the **Create** button to open the forms to create a new instance.

Choose a subscription, resource group, and region to create the instance in. You also need to reference (or create) a new workspace. A workspace is an instance of Azure Log Analytics. This is a central hub that stores and performs analysis on logs. It provides the storage and processing backend for Application Insights:

PROJECT DETAILS

Select a subscription to manage deployed resources and costs. Use resource groups like folders to organize and manage all your resources.

Subscription *   ○   PAYG Dev/Test

        Resource Group *   ○   (New) RG-MTAEDA
                                 Create new

INSTANCE DETAILS

Name *   ○   mtaeda_appinsights

Region *   ○   (US) Central US

Resource Mode *   ○   Classic   **Workspace-based**

WORKSPACE DETAILS

Subscription *   ○   PAYG Dev/Test

        *Log Analytics Workspace   ○   (new) DefaultWorkspace-41ceb1e0-3e91-48d6-9d38-abea77d4a507-CUS ...

Figure 7.8 – Creating a new Application Insights instance in Azure

Select **Review + create** and then confirm by clicking **Create**.

After a few minutes, you will be presented with a link to **Go to resource**. Clicking on this should take you to the overview for the new instance of Application Insights.

Copy and save the **Connection String** information, which is needed to connect the Producer and Consumer services:

Figure 7.9 – Copying the connection string for the Application Insights instance

Now that we have an Application Insights instance and connection string, let us start by setting up telemetry for the Producer service.

Streaming telemetry to Application Insights

The steps to add Application Insights to the Producer service are as follows:

1. Open the `producer.csproj` file and add the following two packages:

```
<PackageReference Include="Microsoft
   .ApplicationInsights.AspNetCore" Version="2.21.0" />
<PackageReference Include="Microsoft
   .ApplicationInsights.Kubernetes" Version="2.0.4" />
```

2. In the `Program.cs` file, add the following service additions just before the `build.Build()` command is called:

```
builder.Services.AddApplicationInsightsTelemetry(
    (options) =>
    options.ConnectionString =
    "---insert your connection string from Application
       Insights ---"
);
builder.Services.AddApplicationInsightsKubernetes
   Enricher();
```

3. Rebuild and start the application.

4. After a few seconds, go back to the Application Insights instance in Azure and select the **Live metrics** blade. You will be able to see a single server, which represents the Producer container, streaming classic instrumentation metrics such as CPU, commitment memory, and request counts:

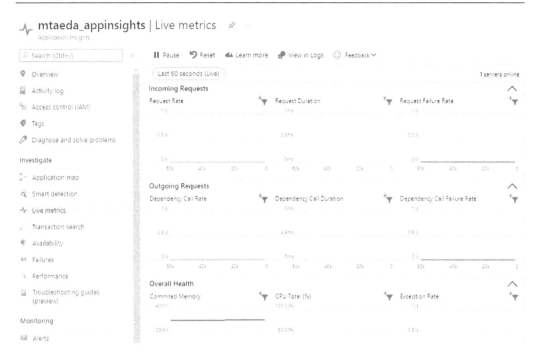

Figure 7.10 – Live instrumented metrics showing one instance of the Producer service

Now, let us follow a similar set of steps to add Application Insights to the consumer service. Be sure to note the subtle differences between this and the producer code:

1. Open the `consumer.csproj` file and add the following two packages:

    ```
    <PackageReference Include="Microsoft
      .ApplicationInsights.WorkerService" Version="2.21.0"
      />
    <PackageReference Include="Microsoft
      .ApplicationInsights.Kubernetes" Version="2.0.4" />
    ```

2. In the `Program.cs` file, add the following to the end of the `CreateHostBuilder()` method. Note that we are extending the existing single code line:

    ```
    }) // End of the ConfigureAppConfiguration parentheses
    .ConfigureServices(
      (config) => config. AddApplicationInsights
        TelemetryWorkerService(
      (options) =>
    options.ConnectionString =
    ```

```
"---insert your connection string from Application
    Insights ---"
    ).AddApplicationInsightsKubernetesEnricher());
```

3. Rebuild and start the application.

4. After a few more seconds, go back to the Application Insights instance in Azure and select the **Live metrics** blade. You will now be able to see two servers, which represent the Producer and the Consumer containers. We now have aggregated visibility into both classic instrumentation metrics:

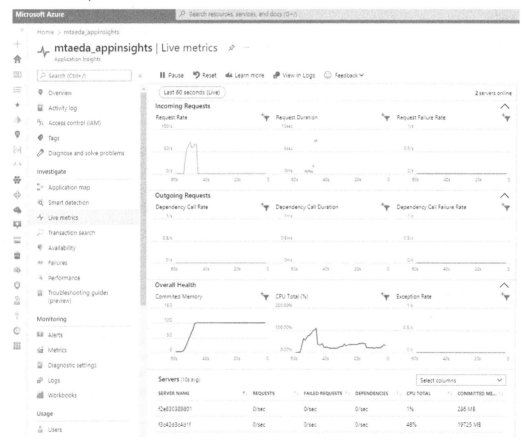

Figure 7.11 – Live instrumented metrics showing one instance each for the Producer and Consumer services

By enabling telemetry, we have configured the connection string and communication channel for the Applications Insights instance. In the code, a Kubernetes Enricher service was also added, which modifies logging information to make it more relevant to hosted containers. For example, the metrics show against a container's instance ID as the server's name, and not the underlying node's hostname. Thanks to this, we can plan to scale our application services in a Kubernetes environment and get the information necessary to view container-specific metrics, as well as aggregated service-level metrics.

While it is great that we now have visibility into more detailed metrics for the services, our primary objective is to aggregate logs. This configuration is a precursor to transmitting direct logging from our services to Application Insights.

Streaming logs to Application Insights

Logging to Application Insights follows the standard `ILogger` pattern for .NET. In setting up the telemetry for metrics, it automatically configured the `ILogger` service to include the Application Insights instance as a destination. All we must do is get access to an `ILogger` service in the code.

Follow these steps for the producer service:

1. In the `Program.cs` file, after we initialize app from `builder`, add the following code:

    ```
    var app = builder.Build();
    // Add the following lines…
    using var scope = app.Services.CreateScope();
    var logger = scope.ServiceProvider
      .GetRequiredService<ILogger<Program>>();
    ```

2. Within the `/send` endpoint code, add a call to the logger right after we set the topic for the producer service:

    ```
    await svc.SetTopic(ProducerTopic);
    logger.LogDebug("Producer service set the topic to
      {topic}", ProducerTopic);
    ```

3. In the `appsettings.json` file, configure the logging level for Application Insights to capture:

    ```
    "Logging": {
        "ApplicationInsights": {
          "LogLevel": {
            "Default": "Trace",
            "Microsoft": "Error"
          }
    ```

```
        }
    }
```

4. Rebuild and start the application.

5. Make a POST call to the /send endpoint of the Producer service.

6. Go back to the Application Insights instance in Azure and select the **Transaction search** blade. Search for topic:

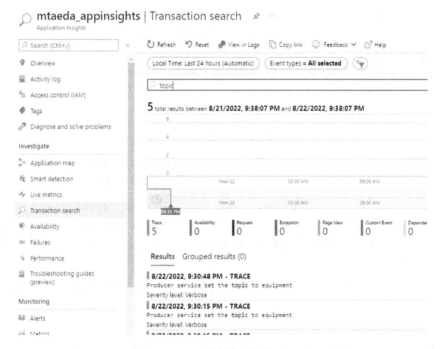

Figure 7.12 – Transaction search to display our log messages

7. We have successfully written a log message to Application Insights. Clicking on one of the trace messages will reveal an **End-to-end transaction details** pane:

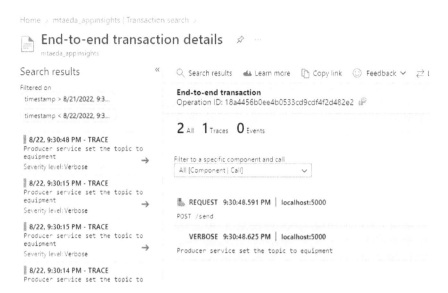

Figure 7.13 – End-to-end transaction details pane

Without adding any further identifiers, Application Insights was able to correlate together the original request and the downstream log message in a linked trace.

This automatic linking of individual logs to produce a trace does not always work when we span across multiple services and event streams. For this, we need to better understand what correlation and causation identifiers are and how we use them in our logs to produce effective traces.

We will explore these concepts further before implementing logging in the consumer service.

Correlation and causation

Correlation identifiers have been critical to monolithic application logging for as long as I can remember. This involves generating a **unique identifier**, which is then passed from function to function and included in any log outputs. That way, you can query lots of log entries using a single common identifier to correlate them together. This forms a trace within the logs so that you can observe a connected pattern of messaging from within the application code.

As we shift to microservices, this challenge goes beyond a single application process. Most commonly, correlation identifiers are propagated from one service to the next using **HTTP for correlation**.

When we move to event-driven architecture, we introduce a few new challenges to this correlation pattern.

Correlation identifiers need to persist across events by being included in the event message from the producer. Then, any consumer can reuse the correlation identifier in its own log output.

While that solves the cross-service trace challenge, we should remember that an event could have any number of subscribers, and they will all persist with the same correlation identifier. This means that when we query by that identifier, the resulting logs could be a large *ball of mud*, lacking enough granular trace threads to be efficiently interrogated.

To address the *ball of mud*, we can introduce **causation** identifiers. A causation identifier is the unique identifier of the preceding event that caused a method to fire. It very simply chains together the preceding event with the current event, so we know what caused what.

Application Insights supports this pattern, which is known as **distributed tracing**. It uses slightly different names for the identifiers:

- Instead of `Log Item Id`, we use `Id`.
- Instead of `Correlation Id`, we use `operation_Id`.
- Instead of `Causation Id`, we use `operation_ParentId`.

The following is an example of these identifiers in action. Multiple consumers have been added to illustrate how the correlation/causation pattern works:

Service	Action	Log Item Id (Id)	Correlation Id (operation_Id)	Causation Id (operation_ParentId)
Producer	POST/SEND	1a	1a	null
Kafka	Message stored	2b	1a	1a
Consumer 1	Message consumed	3c	1a	2b
Consumer 2	Message consumed	4d	1a	2b

Table 7.1 – Example of the correlation and causation identifier pattern

To chain our logs together using correlation and causation identifiers, we need to modify the producer to include these identifiers in the message payload. Previously, we noted that the open source **CloudEvent** specification would be the format ultimately used for event messages in the MTAEDA application. Still, we will use a simple structure as a placeholder. This will consist of only the message string and the identifiers we want to include in the event.

Producing logging and an event payload with correlation and causation IDs

In the producer project, open the `producerService.cs` file and add a definition for the `CloudEvent` structure:

```
internal class CloudEvent : ISerializer<CloudEvent>
    {
        public string? Id { get; set; }
        public string? OperationId { get; set; }
        public string? OperationParentId { get; set; }
        public string? Message { get; set; }

        public byte[] Serialize(CloudEvent data,
           SerializationContext context)
        {
            return Encoding.UTF8.GetBytes(string.Join("|",
              data.Id, data.OperationId,
                data.OperationParentId, data.Message));
        }
    }
```

So, since this class can be used as a Kafka event payload, we must implement the `Confluent.Kakfa.ISerializer` interface. This concatenates the properties with a pipe separator and encodes them in `UTF-8` to a `byte` array.

Before we start to generate or cascade identifiers, first, we want to be compliant with HTTP for correlation. This means that even our producer's `Send()` method should check the `HttpContext` headers for the existence of `Request-Id`, which would indicate that a *pre-existing* correlation ID has already been generated by an upstream service.

We must get access to the `TelemetryClient` service so that if we are provided with a pre-existing correlation ID, this is used in all Application Insight logs for this request.

In the `Program.cs` file, add this line to get the `TelemetryClient` service:

```
var telemetryClient = scope.ServiceProvider
   .GetRequiredService<TelemetryClient>();
```

The causation ID for the event message is the ID of the current request. We get this from `System.Diagnostics.Activity`.

The correlation ID is either `Request-Id` from the **HTTP context header** if we are persisting a *pre-existing* upstream correlation ID, or otherwise, it will be set to the same value as the ID for the current request.

Finally, regardless of what determined the value of the correlation ID, we must set the current telemetry client operation ID to the correlation ID so that it is present in any logging.

It is important we do this at the very start of the `/send` method so that any logs emitted already have the right correlation information and do not generate a new thread of information:

```
var causationId = System.Diagnostics.Activity.Current.TraceId.
    ToString();
var correlationId = http.Request.Headers.RequestId
    .FirstOrDefault() ?? causationId;
telemetryClient.Context.Operation.Id = correlationId;
```

We need to modify the `IProducer` interface definition for the `Send()` method to allow the correlation and causation IDs to be passed in:

```
internal interface IProducerService
{
    Task Send(string message, string correlationId,
        string causationId);
    Task SetTopic(string topicName);
}
```

With this, we also need to update the `producerService` implementation of `Send()` to accept this ID and construct a `CloudEvent` payload with it:

```
public async Task Send(string message, string
    correlationId, string causationId)
        {
            await Task.Run(() =>
            {
                var messagePacket = new Message<int,
                    CloudEvent>()
                {
                    Key = 1,
```

```
                    Value = new CloudEvent()
                    {
                        Id = Guid.NewGuid().ToString(),
                        OperationId = correlationId,
                        OperationParentId = causationId,
                        Message = message
                    }
                };
                _producer.ProduceAsync(Topic,
                  messagePacket, CancellationToken.None);
                _logger.LogInformation("Producer sent the
                  message: " + message + " - and with an Id
                    of: " + messagePacket.Value.Id);
            });
        }
```

Notice how we set the identifiers for the event payload. We generate a new `Guid` for the event's stored ID. We set the correlation ID to `OperationId` and the causation ID to `OperationParentId`.

All that remains for the producer service is to pass the IDs in the `/send` method to the `Send()` call:

```
  await svc.Send(message, correlationId, causationId);
```

Now, we can look at how we address the same challenge in the Consumer service.

Consumer logging with consistent correlation and causation IDs

In the consumer project, add a `CloudEvent` class, equivalent to that added to the producer, in the `consumerService.cs` file:

```
  internal class CloudEvent : IDeserializer<CloudEvent>
      {
          public string? Id { get; set; }
          public string? OperationId { get; set; }
          public string? OperationParentId { get; set; }
          public string? Message { get; set; }
```

```
    public CloudEvent Deserialize(ReadOnlySpan<byte>
      data, bool isNull, SerializationContext context)
    {
        var messageParts = Encoding.UTF8
          .GetString(data).Split("|");
        return new CloudEvent()
        {
            Id = messageParts[0],
            OperationId = messageParts[1],
            OperationParentId = messageParts[2],
            Message = messageParts[3]
        };
    }
}
```

Notice how this class is similar to what was implemented in the producer, except this implements the `Conflient.Kafka.Deserialize` service.

In the `Program.cs` file, we get an instance of the `TelemetryClient` service and pass it into `consumerService` so that we have a logging mechanism:

```
var telemetryClient = scope.ServiceProvider
  .GetRequiredService<TelemetryClient>();
```

The telemetry client is pushed down even further into `MessageRecievedEventHandler`, which ultimately reacts to any new Kafka events published. In its code, we set the operation ID of the client to the operation ID from the event payload. Remember that this is the correlation ID we want to persist through all logs across all services. We also set the parent ID to the ID of the message payload:

```
public async Task Handle(ConsumeResult<int, CloudEvent>
  result, TelemetryClient telemetryClient)
    {
        await Task.Run(() =>
        {
            telemetryClient.Context.Operation.Id =
              result.Message.Value
                .OperationId.ToString();
```

```
telemetryClient.Context.Operation.ParentId
    = result.Message.Value.Id.ToString();
telemetryClient.TrackTrace("Consumer
    reacted to the message: " +
        result.Message.Value.Message);
    });
}
```

Rebuild and start the application, then make a POST call to the /send endpoint of the Producer. Here are the steps for that:

1. Go back to the Application Insights instance in Azure and select the **Transaction search** blade. Clear the search and hit the button to show transactions from the past 24 hours.

2. Sometimes, it can take a few minutes for all logs to be emitted from the services, and received and processed by Application Insights. Eventually, any single /send request should collapse into one trace entry, showing the final Consumer message:

Figure 7.14 – Collapsed trace logs showing the final consumer log message

Clicking on one of the traces will show the full complement of logs that are tied together by the correlation and causation IDs:

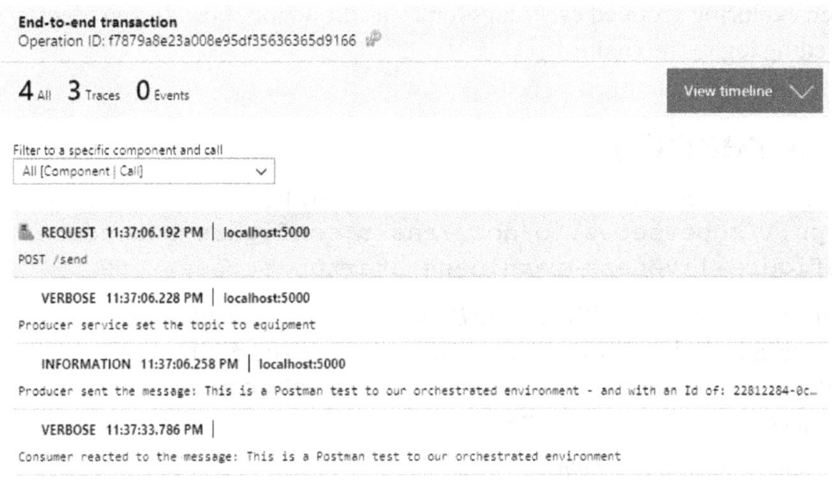

Figure 7.15 – A full end-to-end trace spanning both the Producer and Consumer services

We've done it! We have managed to harness metrics, logs, and traces across distributed services within the event-driven architecture, with some help from Application Insights. This approach is just one of the many you can take to enable observability across your event-driven architecture. All of those approaches will share the same core tenants of aggregation, correlation, and causation.

Summary

In this chapter, we tackled one of the most challenging aspects of effective DevOps for any distributed application – observability. We have managed to aggregate information in a highly scalable way without sacrificing traceability.

We learned about the three pillars of observability, how to create liveness and readiness endpoints, how to aggregate logs, and how to draw effective relationships in logs to trace events across different service processes.

In the next chapter, we will start to look at the build pipelines and integrated testing of the application.

Questions

1. If a liveness endpoint fails for a Pod in a Kubernetes cluster, what is the effect?
2. If a readiness endpoint fails for a Pod in a Kubernetes cluster, what is the effect?
3. When exploring aggregated logs across services, how are all logs following a specific triggering action grouped?

4. When exploring grouped event logs for a specific action, how do you identify a preceding log in the chain?

Further reading

- *Configure Liveness, Readiness and Startup Probes* by `kubernetes.io`, available at `https://kubernetes.io/docs/tasks/configure-pod-container/configure-liveness-readiness-startup-probes/`

- *Monitoring Health of ASP.NET Core Background Services with TCP Probes on Kubernetes* by Rahul Rai, available at `https://thecloudblog.net/post/monitoring-health-of-aspnet-core-background-services-with-tcp-probes-on-kubernetes/`

- *Microsoft Application Insights for Kubernetes* by the Microsoft Open Source community, available at `https://github.com/microsoft/ApplicationInsights-Kubernetes`

- *Logging in .NET Core and ASP.NET Core* by Microsoft, available at `https://learn.microsoft.com/en-us/aspnet/core/fundamentals/logging/?view=aspnetcore-7.0`

- HTTP Correlation protocol implementation in a .NET runtime by the Microsoft Open Source community, available at `https://github.com/dotnet/runtime/blob/main/src/libraries/System.Diagnostics.DiagnosticSource/src/HttpCorrelationProtocol.md`

- Application Insights as a data model for telemetry correlation, by Microsoft, available at `https://docs.microsoft.com/en-us/azure/azure-monitor/app/correlation#example`

8
CI/CD Pipelines and Integrated Testing

Continuous integration (**CI**) and **continuous delivery** (**CD**) are two concepts that have been at the heart of software development for many years. The notion that faster compilation and faster delivery of code to environments brings faster feedback has allowed developers to receive and iterate on positive or constructive feedback. The end goal of this process is to release a better software product.

The discovery and usage of design patterns related to Agile software development do not have to be onerous. There are general constructs that help developers to better build and release their products based on the workflow that melds best with their team setup. Platforms such as GitHub take things a step further and can provide solutions to common build and release situations out of the box.

Throughout this chapter, we will explore different patterns, build pipelines using GitHub Actions, and incorporate integration testing that will help safeguard against introducing bugs that would break existing functionality.

By the end of this chapter, you will be able to do the following:

- Evaluate common CI/CD patterns based on Git repo configuration and popular workflow patterns
- Implement CI/CD patterns using **GitHub Actions** to enable code builds, image builds, and deployments to various environments

- Learn about integration test approaches and implement testing for baseline feature validation, as well as regression testing

> **Important note**
> The links to all the white papers and other sources mentioned in this chapter can be found in the *Further reading* section toward the end of this chapter.

Reviewing common CI/CD patterns

As with any engineering discipline, you can generally find patterns that help implement a specific strategy or outcome. This is also true when you are dealing with CI/CD pipelines. Branching strategies for your repository can be loosely or tightly coupled to the pattern you decide to go with but are not in themselves CI/CD patterns.

As we are using Git as our source control management system, the notion is that you will generally be developing in a branch, pushing those changes remotely, and ultimately, creating a pull request to merge your code into a branch suitable for building a deployable artifact.

However, there are some patterns that, when automated, will give you an advantage, especially if multiple environments or artifacts are used.

Environment-based

Using an environment-based approach to software deployment is typical of most development life cycles. Two patterns are commonly leveraged to facilitate multi-environment deployments:

1. The first is to have all environments captured in one pipeline, which controls whether each environment can be either rolled forward or rolled back.

2. The second is to have one pipeline that can be executed for any environment.

As an example, you may wish to have a standardized pipeline set up for any environment to ensure every deployment is consistent.

Figure 8.1 illustrates a singular pipeline for any environment approach:

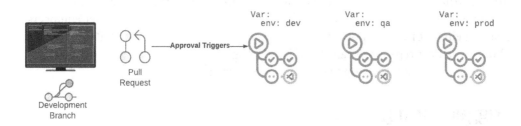

Figure 8.1 – Singular pipeline for any environment

Depending on your workflow, you may want to have one pipeline that will go from build to production. This gives a single pane of glass view into the life cycle of a specific artifact or set of artifacts. *Figure 8.2* shows a singular pipeline for all environment approaches:

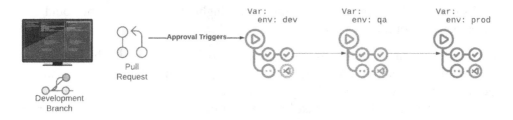

Figure 8.2 – A singular pipeline that contains all environments

Depending on the platform, you may be able to use a consistent environment template and insert that into a comprehensive pipeline that would flow from environment to environment, using the same template for each.

For example, GitHub allows users to create reusable workflows that can be imported and used by several developers. Azure Pipelines also allows for reusable templates in the form of nested YAML template files.

With some examples of environment-based pipelines now covered, we can look at the next pattern commonly found in CI/CD pipelines, which is **artifact management**.

Artifact management

In the world of software engineering, an artifact can mean many things to many different people. Some engineers may see an artifact as an executable program, an installation package, or a series of libraries. Others may see an artifact as a collection of HTML, CSS, and JavaScript files that make up a static web application. Regardless of the type, an artifact in CI/CD terms is anything that can be deployed to one or more environments to provide functionality.

Storing artifacts can be done in several ways. In most cases, artifacts are stored in a location that is accessible to agents of deployment, whether those are other engineers, other pipelines, or even repositories for targets such as Docker images. GitHub has a built-in artifact repository called GitHub Packages, which we'll take a closer look at when we talk about setting up the CD pipeline later in this chapter.

Triggers and gating

To leverage the benefits of a CI/CD platform, it's normally expected that one or more of the branches in a repository has been set up to automatically build whenever a commit is pushed. This CI pipeline invocation is **triggered** by a commit being pushed remotely. Deployments can also be triggered based on successful CI builds; however, the CD pipeline automatically moving code to an environment is typically relegated to much lower environments, where breaking changes are more tolerated and expected. In the majority of cases, deploying code to environments higher than development would require some sort of review.

Inevitably, there will be conditions that require sign-off before releasing the code to an environment. For example, there is typically some type of sign-off needed, especially in large organizations, whenever code is pushed to production. For many teams, this sign-off is a manual process that could take minutes, hours, or even days. A common friction point with developers is the time it takes to resolve manual interventions, including sign-off activities. The conditions that require some sort of intervention are often known as **gates**.

In all modern CI/CD platforms, there are constructs in place to accommodate most conditions that would require manual intervention. While some systems may have more robust implementations – allowing approval-based workflows as well as automated gates based on work items or test results – you can find implementations to suit your needs in the platform of your choosing.

Feature flags

Often seen as an advanced pattern, the use of **feature flags** helps enable continuous delivery by hiding functionality that may not be fully ready for production behind a flag, or rolling back features if there are issues in production. You may decide that a certain group of users can have early access to certain upcoming features to help get feedback directly from them. This practice is known as using a **canary group**, where a small collective of users can test new features and provide feedback, especially in the event of something not working as intended.

Implementing feature flags allows you to shield users from unfinished features but allow the code to move to production. This sets the stage for early testing of new features side-by-side with existing code.

Enabling GitHub Actions for CI/CD implementation

In this section, we'll be walking through how to set up the CI and CD pipelines for the Equipment domain. Once you've completed this example, you will be able to create the same pipelines for all other domain projects.

GitHub Actions for continuous integration

For the continuous integration setup, we will be starting with a basic template meant to build the source code, as well as run any applicable unit tests. Once complete, any required Docker images will be stored in GitHub Packages:

1. Clicking on **Configure** will spin up a new CI pipeline template based on the output you selected. *Figure 8.3* illustrates the pipeline template we will use in this example:

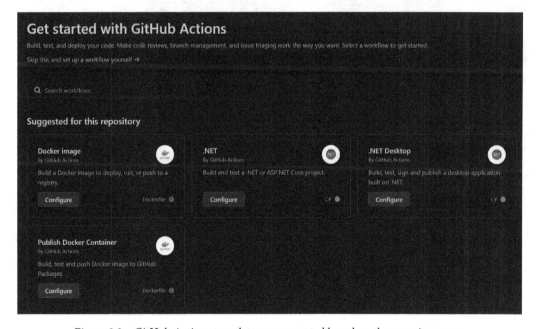

Figure 8.3 – GitHub Actions templates, as suggested based on the repository type

2. *Figure 8.4* shows the YAML pipeline created for you by GitHub when you click on the **Configure** button:

Figure 8.4 – Sample pipeline generated by GitHub when you click Configure

3. We'll also want to set up Packages to be the default artifact repository for this pipeline. You can do this by going back to **Actions** on the main page, selecting the **Publish Docker Container** template, and clicking **Configure**.

Figure 8.5 shows the template that's generated once you click **Configure**:

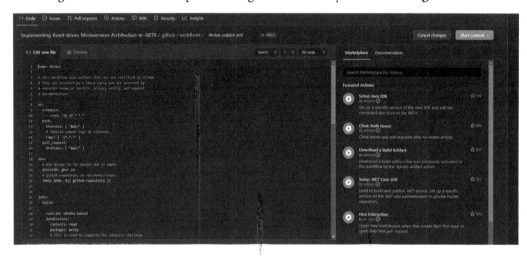

Figure 8.5 – Sample pipeline generated by GitHub for publishing to a Docker registry

With our continuous integration pipeline now configured, we can move on to setting up the second part of our CI/CD combination – the CD pipeline.

GitHub Actions for continuous deployment

We will follow a similar path as we did for the CI pipeline:

1. To create the CD pipeline, we must scroll down to **Deployment** and click on **View all** to see all the deployment flows, as shown in *Figure 8.6*:

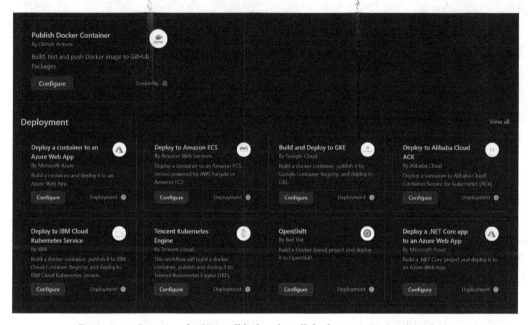

Figure 8.6 – Locating the View all link to list all deployment Action templates

2. From there, we can type `azure` into the search bar to look up specific Azure deployments. For our purposes, we'll want to look for the **Deploy to AKS** template, as shown in *Figure 8.7*, since our deployment target will be **Azure Kubernetes Service**:

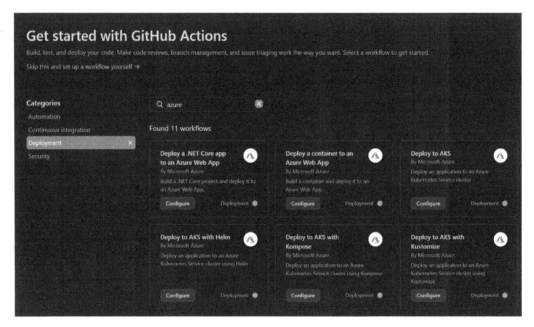

Figure 8.7 – Searching for and using the Deploy to AKS template

With our CI and CD pipelines created, we can deploy the Equipment domain's services. Our next focus will be to work out how integration testing can be added to our pipelines.

Choosing an Integration Test Suite Methodology

Just as there are preferred methods for identifying and writing unit tests for your code, there are similar methods for identifying and writing integration tests. As a general rule, integration tests are tests involving real instances of services, components, or other dependencies as opposed to mocks or stubs, which are often found in unit testing.

Several methodologies are considered standard when it comes to integration testing (source: `https://softwaretestingfundamentals.com/integration-testing/`). These include the following:

- **Big-bang**: All components are combined together and tested at the same time
- **Top-down**: Top-level units are tested first, then lower-level units are tested (that is, testing an API route and then testing the components that the route relies on)
- **Bottom-up**: Lower-level units are tested first, then top-level units are tested (that is, testing component interaction before testing an API route that relies on those interactions)

- **Sandwich/hybrid**: An intentional mix of top-down and bottom-up methodologies

These approaches offer specific benefits, depending on how you wish to validate your code's functionality. They can also support scenarios in which the order of operations is a deciding factor.

Two ways in which you can incorporate integration testing methodologies into your pipeline are through direct deployment or direct consumption. Let's examine some characteristics of each method.

With direct deployment, you can deploy your code, as well as the code or components that your code is dependent upon. For example, each domain pipeline will build a Docker image and publish it to GitHub Packages, making it available to others for consumption. You could then create a Docker Compose file that pulls down the images of the services upon which your code depends and run them at the same time as your code. From there, executing a test script or series of scripts will cause your code to interact with the freshly deployed dependencies, thus yielding a true integration test.

As for direct consumption, this method assumes the dependencies you have are already deployed in some type of integration environment, where the names and locations of the services are well-known. You would then deploy a copy of your code and execute test scripts against it with the understanding that previously deployed services would be in place to interact with your code. This approach reduces the maintenance on your side, only requiring you to publish and deploy your components.

Having talked through examples of integration testing, including how best to work those into our pipelines, we can now look at additional ways of ensuring quality and system resilience.

Summary

In this chapter, we covered some fundamental lessons around the CI and CD of our domain services. We examined some common patterns, such as environment-based deployments, using artifact repositories to store build artifacts for later use, and more advanced concepts such as triggers, gating, and feature flags. We explored how to set up initial CI and CD pipelines using GitHub Actions, including the use of GitHub Packages for storing our artifacts once the builds are complete. We rounded out our pipeline discussion by reviewing some common types of integration testing, and how those tests could be added to our pipelines to further automate our testing and ensure quality is baked into our pipelines.

In the next chapter, we will be taking our testing game even further by incorporating fault injection and chaos testing into our pipelines. This is just one aspect that can help us identify issues with application resiliency and reliability, which is critical when dealing with an event-driven platform that is intended to be highly scalable.

Questions

1. When designing CI/CD pipelines, is it important to make sure all of the patterns discussed are implemented? Why or why not?

2. Is there a case where implementing feature flags would not be beneficial?

3. What are some examples of gates within a software development life cycle that could be automated to help reduce friction and increase time to market?

4. Is there a best practice or preferred methodology for implementing integration tests in pipelines?

Further reading

- **Use continuous integration**: `https://docs.microsoft.com/en-us/devops/develop/what-is-continuous-integration`

- **What is continuous delivery?**: `https://docs.microsoft.com/en-us/devops/deliver/what-is-continuous-delivery`

- **What Are Feature Flags?** `https://azure.microsoft.com/en-us/overview/continuous-delivery-vs-continuous-deployment/`

- **Using environments for deployment**: `https://docs.github.com/en/actions/deployment/targeting-different-environments/using-environments-for-deployment`

- **Reusing workflows**: `https://docs.github.com/en/actions/using-workflows/reusing-workflows`

- **Reviewing deployments**: `https://docs.github.com/en/actions/managing-workflow-runs/reviewing-deployments`

- **Adding approval workflow to your GitHub Action**: `https://timheuer.com/blog/add-approval-workflow-to-github-actions/`

- **Approval Workflows With Github Actions**: `https://www.aaron-powell.com/posts/2020-03-23-approval-workflows-with-github-actions/`

- **Feature Toggles (aka Feature Flags)**: `https://martinfowler.com/articles/feature-toggles.html`

- **Feature flags**: `https://www.optimizely.com/optimization-glossary/feature-flags/`

- **What Are Feature Flags?** `https://launchdarkly.com/blog/what-are-feature-flags/`

- **Integration Testing**: `https://softwaretestingfundamentals.com/integration-testing/`

9
Fault Injection and Chaos Testing

Just as software development practices have evolved over the past few decades, so has software testing. With the added draw of paradigms such as **Infrastructure as Code (IaC)** and **Configuration as Code (CaC)**, more and more pieces of an application are now stored in code and can be tested in a variety of ways. An added benefit of leveraging cloud services to host portions of an application is the ability to test, in isolation or production, how the application will respond in the event of a cloud platform failure.

At times, anything from virtual machine services to serverless functions to authentication platforms can become unavailable. Having the ability to see how your application reacts in the face of unexpected outages is something that is often overlooked when dealing with applications running in an on-premises environment. Being able to anticipate, script, test, and resolve issues based on your understanding of potential issues brings more confidence in your development life cycle, as well as more confidence in the stability of your application.

We'll cover the following topics in this chapter:

- Fault tolerance and fault injection
- Chaos testing
- Implementing fault injection and chaos tests

By the end of this chapter, you will be able to do the following:

- Understand what fault injection is and how it relates to the overall tenet of fault tolerance

- Understand chaos testing and how it can augment fault injection testing

- Implement fault injection and chaos tests in your pipelines

Technical requirements

You can find all the code examples for this chapter in this chapter's folder on GitHub at `https://github.com/PacktPublishing/Implementing-Event-Driven-Microservices-Architecture-in-.NET-7/tree/main/chapter09` and `https://github.com/PacktPublishing/Implementing-Event-Driven-Microservices-Architecture-in-.NET-7/tree/main/src`.

> **Important note**
> The links to all the white papers and other sources mentioned in this chapter can be found in the *Further reading* section toward the end of this chapter.

Fault tolerance and fault injection

The concept of **fault tolerance** – that is, the ability of an application, platform, or runtime to tolerate a systemic fault – by itself seems a simple enough concept to grasp. After all, you would expect an application to be able to gracefully recover if certain services were not available. In many cases, though, applications have been written with an understanding that the underlying infrastructure that hosts it is always available unless a catastrophic event occurs. While this reliability may be built into on-premises data centers and rarely challenged, the same assumption does not hold for cloud platforms, services, and components.

Though cloud platforms will offer certain **Service-Level Agreements** (**SLAs**) for uptime on some cloud services, there is always the possibility of a service-level or region-level outage that can come with no warning and vary widely in impact. Therefore, it is important to keep these types of outages in mind when you are developing an application for the cloud.

Researching the status pages of cloud providers can provide great insight into what services have been affected by outages, how frequently they are affected, and what remediations were taken to fix the issues. Using this information can help you to evaluate which faults or outages are most likely to occur, and how you can better prepare your application to deal with them.

Anticipating and tolerating faults

Throughout this book, we have looked at decoupled and asynchronous patterns for communication between components in our application. These patterns are not fault-proof, but they do allow us to handle unexpected interruptions in a less disruptive manner. As a part of application design, it's common to look at what the application will be hosted on, the requirements for the application being available, and how the application can maintain functionality even when the unthinkable occurs – a total outage of a data center or cloud region. Part of that design process will take you through the various *what-if* scenarios and lead you to identify faults that may occur based on how and where you host the application. We will discuss the impact resiliency has on the use of design patterns for scalability in more detail in *Chapter 10, Modern Design Patterns for Scalability*.

Using your faults against you

One interesting paradigm that has emerged from the testing community is the notion of purposely introducing errors, exceptions, and faults during testing to measure the application's reaction. **Fault injection** is the practice of intentionally causing conditions that will result in an underlying operating system, hosted service, or runtime failing. This practice has long been used by engineers for testing physical hardware by affecting circuits, chips, wires, or power supplies to determine how the device will respond. This same mantra eventually transitioned into the software world.

Using this approach, testers can simulate different types of environmental conditions that may cause an application to perform poorly or stop working altogether. While this behavior is anticipated, what the application does to compensate for that is what is measured.

Tests that perform fault injections can be written in various ways, from single scripts to entire libraries that focus on a litany of different conditions or errors. One approach that has gained popularity is that of **chaos engineering**, which results in the ability to run **chaos testing** against your application to break the underlying platform and uncover areas for improvement.

Chaos testing

While **chaos testing** involves executing tests designed to break services your application depends on, the overall discipline of **chaos engineering** was first established in 2010 by engineers at **Netflix**. The primary purpose of this type of engineering was to test how their services and applications behaved under extreme circumstances. In 2012, Netflix open sourced **Chaos Monkey**, which was their internal chaos testing platform. This opened the doors for other companies and engineers to be able to leverage and modify that suite and adapt it to their applications.

The order of chaos engineering

Chaos engineering follows a simple set of principles, allowing there to be order within the chaos, so to speak. The three main principles of chaos engineering, according to **Gremlin**, are as follows:

1. **Plan an experiment**: This step starts with a hypothesis you formulate based on how your application would respond to a problem or outage. This can depend heavily on your definition of a stable or steady state – what is considered *normal* for the application. Understanding and documenting non-functional requirements, including API response times, page load times, and other **Key Performance Indicators** (**KPIs**), go a long way in determining what is *normal* versus what is problematic. Also, consider the information you gathered from your cloud provider on service outages, as this will help you recreate real-world outage situations.

2. **Contain the blast radius**: This step involves creating a test that simulates this problem or outage in the smallest possible footprint possible. For example, if the problem is data loss, you may start by experimenting with deleting one row in a database table, or one document within a **NoSQL** data store.

3. **Scale or squash**: This final step is where you determine whether the test proved out. If you found an underlying issue, then you have a path forward to improve how your application responds to that problem. If not, continue to scale up the test, increasing the blast radius until you are at full production capacity. If you uncover an issue when scaling your activity up, great – if not, also great!

Using these principles, you can start formulating different failure scenarios that your application might experience, and from there put together hypotheses about how the application should behave given the failure type.

Implementing fault injection and chaos tests

Suites such as Gremlin aim to make the adoption of chaos engineering easier to move toward since the Chaos Monkey suite popularized by Netflix can have an aggressive learning curve. For **Kubernetes**-based applications, a popular framework for chaos testing is **Chaos Mesh**. While it has not been around as long as suites such as Chaos Monkey, Chaos Mesh is a solid choice for implementing chaos experiments in a uniform way against a modular orchestration engine. Chaos Mesh is also an incubating project with the **Cloud Native Computing Foundation** (**CNCF**), which exposes it to a larger open source community for use.

It's interesting to note that **Azure Chaos Studio**, which we will be using to set up some baseline experiments, relies on Chaos Mesh behind the scenes to orchestrate and run experiments against Kubernetes-based targets. It also allows for agent-based installations on virtual machines, as well as virtual machine scale sets and other resources, giving you more control over where and how experiments can be run. We will start by diving into Chaos Mesh before moving on to orchestrating experiments using Azure Chaos Studio.

Starting with Chaos Mesh

First, we will look at installing and using Chaos Mesh to get a feel for how the experiments can be structured to impact services, namespaces, and even hosts within the Kubernetes cluster. It can be installed in any valid Kubernetes environment, from **minikube** to **Docker Desktop** to cloud-hosted Kubernetes clusters. Each experiment is created as a **Custom Resource Definition** (**CRD**) within the cluster, which allows specific .yaml files of every experiment you want to run to be captured; they can then be applied during a pipeline or test run.

If you happen to be using a **Linux** distribution, **Windows** subsystem for Linux, or **macOS**, the one-liner provided on the Chaos Mesh home page can get you started with the installation process:

```
curl -sSL https://mirrors.chaos-mesh.org/v2.2.2/install.sh |
   bash
```

Doing so will result in Chaos Mesh being created in the cluster you have selected. If you do not wish to install this via the command line, you can also install Chaos Mesh using the **Helm** chart by adding the Helm repository to your cache and installing it from there. Once the installation is complete, you should inspect the namespace and resources within it. Three services will be installed:

- **Chaos-daemon**: This accepts and executes chaos experiments, including injecting any of the supported experiment types.

- **Chaos-dashboard**: This allows you to manage the mesh using a web-based UI.

- **Chaos-mesh-controller-manager**: This is the control plane for the mesh. It sits between the Kubernetes APIs and the chaos-daemon service, receiving events from the Kubernetes APIs and sending those to the chaos-daemon service for execution.

Using **kubectl**, you can access the Chaos Mesh dashboard by port-forwarding:

```
Kubectl port-forward -n chaos-mesh svc/chaos-dashboard
    9000:2333
```

Follow these steps to configure Chaos Mesh:

1. Upon connecting for the first time, you will be asked to enter a **Role-Based Access Control** (**RBAC**) token to continue, as shown in *Figure 9.1*:

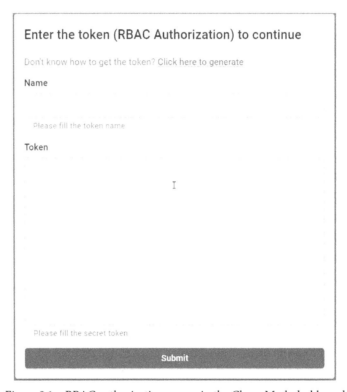

Figure 9.1 – RBAC authorization screen in the Chaos Mesh dashboard

2. If you do not have a token, click on the **Click here to generate** link; instructions will be provided to you. A new window will open, as shown in *Figure 9.2*. The default settings are to start with a scoped token, allowing you to choose the namespace and the role with which to connect. For this example, we will be using a cluster-scoped RBAC token to allow for maximum visibility and management. It is not normally considered a best practice to generate these tokens without scoping them down to at least a namespace level:

Figure 9.2 – Token generator screen in the Chaos Mesh dashboard

3. Be sure to follow the instructions as shown in the dialog, and copy the token name as well as the token itself (shown upon running the `kubectl describe` command). You will need to click on **Close** to close this dialog and return to the RBAC screen shown in *Figure 9.1*. Enter the token name and the token value in the appropriate boxes, then click **Submit**. This should reload the dashboard for you, showing a home page similar to what is displayed in *Figure 9.3*:

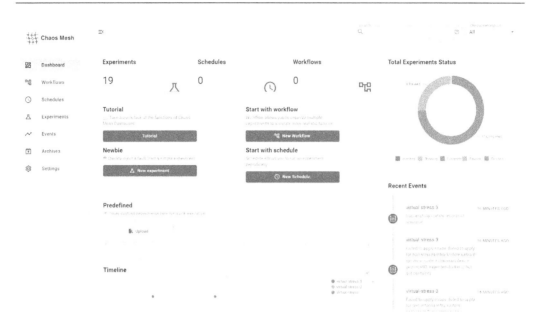

Figure 9.3 – Sample home page display of the Chaos Mesh dashboard

4. From here, we can start by clicking on the **New experiment** button to create a new basic chaos experiment. This will load a new screen similar to what is shown in *Figure 9.4*. **Kubernetes** will be selected by default, and you can click on **Stress Test** to create a new stress test experiment in the cluster. Options will appear so that you can fill in the number of CPU workers, the percentage of CPU load to simulate, and additional configuration options. Fill in the values, as shown in *Figure 9.4*, and scroll down the form until you see a **Submit** button. Click that to set the values for the CPU stress test and move on to the next section, where you can set up the targets for the experiment:

Figure 9.4 – New experiment options

5. Select the **mtaeda-infra** namespace to target the main infrastructure for Kafka. This will show a listing of pods within that namespace that will be impacted by the experiment. Give the experiment a name and enter the duration of the experiment in minutes. The resulting information will look similar to what's shown in *Figure 9.5*:

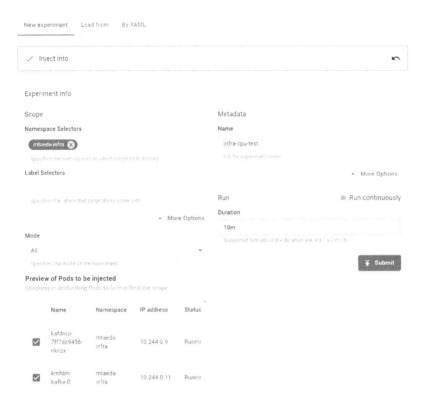

Figure 9.5 – Experiment information required for new chaos experiment

6. Click **Submit** – this will lock in the experiment information. One final **Submit** button will be displayed – upon clicking this button, you will be taken to the **Experiments** tab, which shows the progress of your experiment. Click on the name of your new experiment to be taken to a detailed view, showing any actions issued against the targeted pods, as well as a YAML definition listing the configuration of the experiment. Download this definition and save it as `kafka-cpu-load-test.yaml`.

This definition is important as it does two things:

*   Codifies the experiment to allow for automated injection during a scripted test

*   Outlines the experiment parameters, which will be used when creating an experiment in Azure Chaos Studio

Now that we have taken a quick tour of Chaos Mesh and learned how experiments can be created as well as configured, we are ready to take the next step and delve into Azure Chaos Studio.

Using Azure Chaos Studio

Azure Chaos Studio is a new service provided by Microsoft that allows you to run chaos experiments against a subset of Azure resources. To find it, go to **All Services** in the Azure portal and search for `chaos`. **Chaos Studio** should show up as one of the options, as shown in *Figure 9.6*:

Figure 9.6 – Locating Azure Chaos Studio within the list of Azure services

> **Important note**
>
> At the time of writing, Azure Chaos Studio is in Public Preview. Names, screenshots, and features may be subject to change once the service has reached **General Availability** (**GA**).

Follow these steps to use Azure Chaos Studio:

1. Clicking on the **Chaos Studio** link will take you to the main page of the studio. Here, you will be shown an overview page with options to set up the targets of chaos experiments, as well as the experiments themselves. *Figure 9.7* illustrates the overview page:

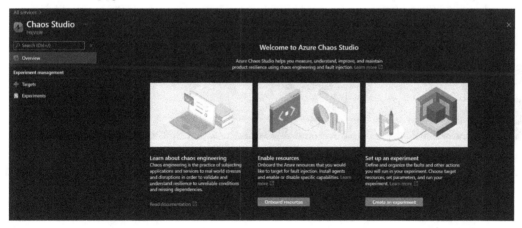

Figure 9.7 – Azure Chaos Studio overview page

2. To start, we will click on **Targets** to see what available Azure resources we can run chaos experiments against. Depending on what you have deployed in your subscription, your list may vary. In this case, we are looking for the virtual machine scale set associated with the AKS cluster we provisioned. *Figure 9.8* illustrates an overview of the supported objects, with the scale set being the first object:

Figure 9.8 – List of applicable targets to enable for chaos experiments

3. Place a check next to the name of the scale set. This should enable the **Enable targets** button. Upon clicking on it, you will be presented with two options, as shown in *Figure 9.9* – one to enable service-direct targets and another to enable agent-based targets. Clicking on either of these options will enable the resource you selected with those chaos experiment targets:

Figure 9.9 – Options for enabling experiment targets for a selected resource

4. Click on the service-direct link first for the scale set and wait for the operation to complete. Once that has finished, click on the agent-based link to install the agent on the machines in the scale set.

5. Next, find the AKS cluster you will be using for the test and perform the same action. You'll notice that only the service-direct option will be enabled. Click on that and wait for the operation to finish. Once that's finished, click on the **Manage actions** link to the right of the row containing the cluster. This will open a new blade, as shown in *Figure 9.10*, cataloging the available service-direct chaos experiment options for the cluster:

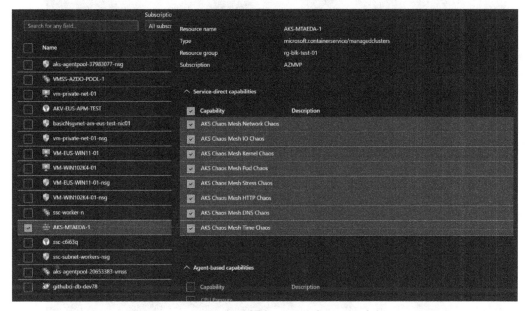

Figure 9.10 – List of available service-direct capabilities

6. This list targets actions that can be performed by making calls to the Kubernetes APIs. For other resource types, the capabilities shown will reflect properties or actions that can be affected by using the **Azure Resource Management API**. When clicking on the **Manage actions** link for the virtual machine scale set, for example, you will be presented with a list of capabilities that can be run directly on a machine, due to the agent being installed as a part of enabling the agent-based experiment types. *Figure 9.11* illustrates examples of the agent-based experiment types that get enabled for a virtual machine or a virtual machine scale set:

Figure 9.11 – List of available agent-based capabilities

7. We can now go to the **Experiments** blade and create a new experiment. Clicking on **Experiments** will bring you to a new screen, which will show you existing experiment definitions (if you have created them before) or will prompt you to create a new one:

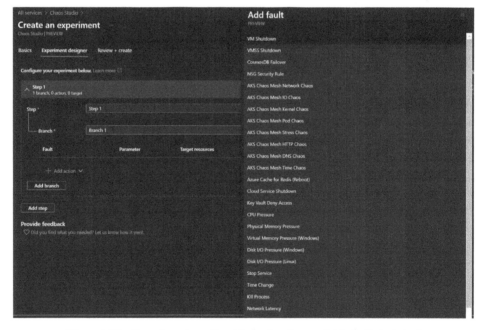

Figure 9.12 – Experiment creation blade showing available faults to inject

8. When editing the fault definition, select **AKS Chaos Mesh Stress Chaos** from the dropdown. You will see two fields to fill in – one for the duration of the experiment and another for the JSON specification, which tells the experiment what to target. The chaos experiment definition file we saved earlier from Chaos Mesh will contain the information we need, but the file is in `.yaml` format. One option is to find an online YAML-to-JSON converter and copy the YAML from the **selector** element down. The rest of the file will not be required.

9. You can also convert the required YAML into JSON manually if you wish. In either event, copy the converted JSON text and paste it into the **jsonSpec** field, as shown in *Figure 9.13*. This will tell the experiment to target the **mtaeda-infra** namespace and run a CPU stress test on pods in that namespace for 10 minutes:

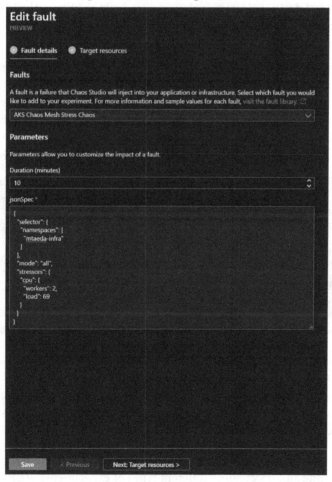

Figure 9.13 – Updating the jsonSpec field with JSON-ified settings from the YAML file

10. The **jsonSpec** field is required – without it, the experiment will not know what to do or what namespace(s) to take action on. If you have not already done so, click on **Next: Target resources >**; you will be presented with a list of AKS clusters you have enabled service-based faults on. Select the appropriate cluster, click **Save**, and return to the experiment editor. Click **Save** at the bottom to be taken back to the main experiment screen. At the top, you will see a **Start** button, which will kick off the experiment. Before you do, some permissions need to be given to the managed identity that was created as a part of the chaos experiment.

11. In the Azure portal, go to the main screen for your AKS cluster. Click on **Access Control (IAM)** and then click on the **Add | Add role assignment** options. For the AKS cluster, you will want to look for the role named **Azure Kubernetes Service RBAC Cluster Admin**. Select this role and click **Next** to choose the identity to assign the role to. Choose **Managed identity** as the type to assign access to, and click on **Select members** to open the **Select managed identities** pane.

12. Under **Managed Identity**, use the dropdown to find **Chaos Experiments**, which should have at least one identity associated with it, as shown in *Figure 9.14*. Click on the identity that appears once you select **Chaos Experiment**, noting that your experiment name should be a part of the identity's name. Click the **Submit** button on the identity pane to select the identity and click on **Next** to advance to the **Review** pane. Click **Review + Assign** and the RBAC permissions will be added:

Figure 9.14 – Adding a role assignment for a managed identity

13. Let's return to the **Experiments** pane and locate the new experiment. Upon clicking it, you will see the **Start** button again at the top of the blade – press this button and confirm that you wish to run the experiment. The experiment will now kick off and run for the duration you specified.

Now that you have started your first successful chaos experiment using Azure Chaos Studio, take some time and try creating different types of experiments. Remember, if you get stuck on the **jsonSpec** field for a particular experiment type, you can always go back to the Chaos Mesh dashboard and create a new experiment there with the same parameters.

This will create the YAML definition for you, which you can tweak to your liking.

Summary

In this chapter, we covered a lot of ground concerning understanding, anticipating, handling, and causing faults. We explored the paradigm of fault tolerance, as well as how to use fault injection to test how well our application tolerates systemic outages. Using different experiments, we've seen how chaos engineering can be used to enable an additional layer of testing, giving us insights into areas of our code that may not handle unexpected problems very well.

As we look forward to *Chapter 10, Modern Design Patterns for Scalability*, the output of these experiments will help us to choose patterns that make sense for our application and the potential traffic it may generate.

Questions

1. What is the difference between a fault and an exception?
2. What are some examples of faults that can be anticipated and handled?
3. Is fault injection a standalone concept? Why or why not?
4. Where does fault injection testing take place? Chaos testing?
5. When designing an experiment for chaos testing, how large of an impact should the test have initially?
6. What types of resources are compatible with Azure Chaos Studio?

Further reading

- **What is fault injection?** `https://www.gremlin.com/blog/what-is-fault-injection/`
- **Azure status history**: `https://status.azure.com/en-us/status/history/`
- **Fault Injection Testing**: `https://microsoft.github.io/code-with-engineering-playbook/automated-testing/fault-injection-testing/`
- **Principles Of Chaos Engineering**: `https://principlesofchaos.org/`
- **The Netflix Simian Army**: `http://techblog.netflix.com/2011/07/netflix-simian-army.html`

- **Chaos Toolkit**: https://github.com/chaostoolkit
- **Chaos engineering**: https://learn.microsoft.com/en-us/azure/architecture/framework/resiliency/chaos-engineering
- **Chaos Engineering: the history, principles, and practice**: https://www.gremlin.com/community/tutorials/chaos-engineering-the-history-principles-and-practice/
- **Chaos Mesh**: https://chaos-mesh.org/
- **Cloud Native Computing Foundation**: https://www.cncf.io/
- **Create a chaos experiment that uses a service-direct fault to fail over an Azure Cosmos DB instance**: https://docs.microsoft.com/en-us/azure/chaos-studio/chaos-studio-tutorial-service-direct-portal
- **Create a chaos experiment that uses a Chaos Mesh fault to kill AKS pods with the Azure portal**: https://docs.microsoft.com/en-us/azure/chaos-studio/chaos-studio-tutorial-aks-portal
- **Azure Chaos Studio**: https://azure.microsoft.com/en-us/services/chaos-studio/
- **Supported resource types and role assignments for Chaos Studio**: https://docs.microsoft.com/en-us/azure/chaos-studio/chaos-studio-fault-providers

Part 3: Testing and Deploying Microservices

This part examines the need for scalability and service resilience within the application, along with implementation details related to elastic and autoscale components and how proper telemetry helps to automatically drive scaling events. The topic of observability is revisited using examples of service discovery and microservice inventories.

This part has the following chapters:

10
Modern Design Patterns for Scalability

In years past, determining how to handle the scalability of your application depended primarily on capacity planning and static resources hosted in a data center. With the rise in popularity of the cloud, more flexible options exist for managing the scalability of your application. That's not to say that capacity planning is not needed – it is still an important part of system architecture design.

How you implement scalability options tends to be different in the cloud, though. Where static resources once were the mainstay, you can now leverage automated scaling options, depending on how you are hosting your application. Cloud services such as managed **Kubernetes**, app services, and managed storage can be individually tuned to provide minimal resource usage when not under heavy load and maximized resource usage when it's needed most.

This chapter will focus on the key metrics to monitor for potential performance issues, and what settings can be fine-tuned to allow for an automatic response to a degradation in performance.

We'll cover the following topics in this chapter:

- Mechanisms for autoscaling
- Implementing autoscaling for Kubernetes services
- Implementing autoscaling for Azure App Service

By the end of this chapter, you will understand the different types of autoscaling based on metrics related to CPU load, **input/output** (**I/O**) load, request load, and memory-intensive operations. You will also know how to implement auto-scaling features for services deployed to Kubernetes and how to implement auto-scaling features for services deployed to Azure Web Apps or Azure API Apps.

> **Important note**
> The links to all the white papers and other sources mentioned in this chapter can be found in the *Further reading* section toward the end of this chapter.

Mechanisms for autoscaling

There are four primary indicators to watch during the operation of an application that can indicate potential issues with maintaining uptime or responsiveness SLAs. These include CPU load, I/O load (often seen as disk pressure in Kubernetes), request and network load (often seen as network pressure in Kubernetes), and memory load. Understanding these indicators is essential in helping to prepare you to adjust configuration settings, thus leading to scalable supporting infrastructure. Understanding how your application components affect each of these indicators is important as well.

Compute and CPU load

The amount of CPU that's utilized will vary heavily across physical machines, virtual machines, and even hosted cloud services. With cloud services, even greater amounts of fine-grained adjustments can be made. For example, with **Azure App Service**, the amount of compute initially available to an App Service is tied directly to the App Service Plan that hosts the service. Thresholds can be set to look for sustained CPU activity that lasts longer than 5 to 2- minutes, anomalies in CPU usage (spikes), or even a point-in-time benchmark, such as a set percentage.

Similar rules apply in the Kubernetes world – with a twist. Kubernetes allows you to not only set monitors on the nodes within the cluster (known as **hosts** in the Kubernetes ecosystem) but also on any running pod within that cluster. What's more, you are not restricted by the sum of processing power within your cluster – it can be overallocated, which can lead to a variety of different problems.

We'll look at the disk and I/O load next, which can bring your application to a grinding halt, quite literally in some cases.

Disk and I/O load

Managing expectations around the usage of disk storage, whether through directly attached storage, flash storage, or managed storage options offered by cloud service providers, is generally something that should not be overlooked. If you know that your application requires a high rate of disk reads or disk writes, you will want to spend some time planning how best to implement those requirements. A new feature in **.NET 6** that can also help with certain elements of possible I/O contention is the rewrite of the `FileStream` object, leading to improved performance and a possible reduction in potential I/O contention issues.

In some cases, applications may use a local log file to write out information about the application's execution or any errors it may have encountered. In others, there may be more of a reason to use disk storage to perform data manipulation or read and write a high volume of transactions. In some financial services applications, using storage to read, sort, and write values is intentionally done to avoid running into bottlenecks with available CPU or memory. If the managed storage being used does not meet the I/O requirements for the application, disk contention can occur, and could lead to more serious impacts on the application. Normally, disk configuration is not something that is set up to autoscale from an I/O perspective, but it can help you to adjust other resources that may help compensate.

The next indicator we'll cover is that of network load, where requests and responses can potentially get jammed up if there isn't enough room for all of them to flow through the network.

Request and network load

Another finitely scoped resource is that of network bandwidth, specifically because there is only so much throughput a virtual or physical network can handle. Normally, networking considerations are considered when designing a regional footprint with a cloud provider, where throughput limitations can be examined, and the appropriately sized network pipe can be installed to allow traffic to flow as expected.

Since tweaking network settings is not something that most developers would want to do, or be expected to do, the main way in which network load can be monitored (and reacted to) is through requests per second. If one of your application's components expects to receive a high volume of requests, but also needs a fair amount of compute and/or memory resources to process each request, system performance will slow to a crawl as the component processes each request. Adjusting the network is not an option but setting components up to autoscale in anticipation of this, or even throttling requests to specific components, can help limit the amount of degradation a user might experience while using your application.

We will now round out this list by diving into memory-intensive operations, which can impact a component within an application, an entire application, or even an entire hosted service.

Memory-intensive operations

Managing the performance of memory during your application's lifespan can be a daunting task, especially if resources are limited. Some programming languages can put demands on memory that cause it to become depleted fast. Java, for example, requires that at least two **gigabytes** (**GB**) of RAM are available to a running instance of the **Java Virtual Machine** (**JVM**) executing your code. If you are deploying this into a Kubernetes environment, this can add up quickly and potentially starve other services of RAM.

Understanding how your application uses memory during its execution can greatly assist you in not only optimizing the starting settings for your application but also how scaling can be implemented to help relieve memory pressure. Setting autoscale events to trigger when memory usage starts climbing for components that leverage a large amount of RAM can help spread operations out to handle memory pressure and still give the user a desirable experience.

Now that we have looked into the main indicators that can lead to performance issues, let's talk about how to implement patterns to adjust to and even anticipate greater needs, starting with applications hosted in Kubernetes.

Implementing autoscaling for Kubernetes services

Kubernetes itself is quite a powerful orchestration platform that allows you to control how few or how many system resources a given bit of executable code can have access to. Since the implementation of Kubernetes can be done on-premises, in the cloud on virtual machines, or through a managed service, there are several options for autoscaling configuration. These can range from patterns in Kubernetes itself to certain features of cloud-managed services to third-party plugins that are purpose-built for specific scenarios.

Native Kubernetes options

As an orchestrator, Kubernetes offers a rich ecosystem that allows you to use as little or as much of the cluster's compute power as needed, in a variety of ways. Some features allow you to control how applications can scale out, depending on some of the primary indicators we covered in the previous section. In this section, we will start with a couple of native options that can be used in your cluster, wherever it resides.

Horizontal Pod Autoscalers

Horizontal Pod Autoscalers (HPAs) are constructs within Kubernetes that allow you to specify a target condition or threshold that a workload must hit before the cluster will scale out and create a new pod or set of pods. The common parameters of minimum and maximum nodes are applicable here, giving you the ability to set up some guardrails around how much scaling occurs. The other parameter is the condition the deployment itself is experiencing – whether the event triggering the autoscaling is CPU or memory bound.

While CPU and memory are the two out-of-the-box means for autoscaling, there are other ways you can add metrics to control autoscaling. *Table 10.1* lists other common targets that can be used in tandem to determine how and when autoscaling should kick in:

Metric Type	Description	Example

Resource	Metrics that are directly related to available pod container resources (CPU, memory).	<pre>- type: Resource
		  resource:
		    name: memory
		    target:
		      type: Utilization
		      averageUtilization: 70</pre>

Pods	Metrics related to pods within a deployment are generally measured as an average across pods (packets per second).	<pre>- type: Pods
		  pods:
		    metric:
		      name: packets-per-
		        second
		    target:
		      type: AverageValue
		      averageValue: 2k</pre>

Object	Metrics related to other objects within the cluster that may impact performance (requests per second).	<pre>- type: Object
		  object:
		    metric:
		      name: requests-per-
		        second
		    describedObject:
		      apiVersion:
		        networking.k8s.io/v1
		      kind: Ingress
		      name: main-route
		    target:
		      type: Value
		      value: 8k</pre>

| External | Metrics related to objects outside of the cluster. An example of this might be a hosted queueing service. | ```
- type: External
  external:
    metric:
      name:
        queue_messages_ready
      selector:
        matchLabels:
          queue:
            "worker_tasks"
    target:
      type: AverageValue
      averageValue: 20
``` |
|---|---|---|

Table 10.1 – Kubernetes metric types related to HPAs

The Kubernetes documentation recommends creating custom metrics instead of using external metric types. This is because securing custom metrics internal to a cluster is easier than trying to secure metrics related to an outside-hosted service. It also removes the dependency on platforms outside of the cluster. In some cases, however, creating an external metric type may apply to your application, and its usage should be closely examined during the design process.

ReplicaSets

A less technical way to control how a Kubernetes deployment remains flexible with scaling is to establish a ReplicaSet with a larger maximum pod value. While using this approach does take the automation out of the scaling operation, it does allow for control over the number of pods in the ReplicaSet. For availability purposes, it is generally a good practice to use a ReplicaSet with your deployments to ensure a minimum number of pods is always available. Controlling the scaling aspect manually instead of automatically would be better suited for low-impact workloads where detecting conditions that require more resources is less likely to cause an issue with the application.

Certain elements of the Kubernetes platform allow you to control autoscaling, regardless of where the cluster is hosted. Next, we will take a look at options that are available to you for scaling concerning where the cluster is hosted – in this case, the public cloud.

Cloud platform options

With each of the cloud platforms that offers Kubernetes services, there are several different ways in which scalability can be configured. As our example deals with Azure, we will
```

examine the settings used to configure autoscaling for **Azure Kubernetes Service** (**AKS**). Keep in mind that Amazon and Google may have similar (or different) means of configuration.

Node pool autoscaling

In AKS, there is a concept of node pools – groups of virtual machines that are a part of the Kubernetes cluster, managing system workloads as well as application workloads you deploy to it. A vanilla deployment will set up a default node pool, normally with three nodes that allow for those workloads to be distributed across them. As we mentioned in the previous section, Kubernetes does allow you to over-allocate resources to the cluster, so having a way for the node pool to adjust based on an increase in resource demand seems like a good option to have.

You can set up autoscaling in AKS in a few different ways. From a visual perspective, you can use the Azure portal to configure the autoscaling settings for the node pool. This tends to help people who are not intimately familiar with IaC. The portal allows you to choose between manual or automatic node scaling. With automatic scaling, you are asked to enter a maximum number of nodes that the pool can scale to. This protects you from an autoscaling event that would create dozens of cluster nodes, which could be rather unnecessary and quite expensive.

The preferred approach to configuring autoscaling for node pools is with IaC. This allows the configuration to be centrally managed and repeatable, eliminating the potential for human error. Using **Terraform**, for example, an AKS configuration might look like the following snippet. Some content has been trimmed so that we can focus on the autoscaling settings:

```
resource azurerm_kubernetes_cluster "aks-mtaeda-1" {
    name = "aks-mtaeda-1"
    ...
    default_node_pool {
      name = "defaultpool"
      enable_auto_scaling = true
      max_pods = 100
      min_count = 1
      max_count = 8
      vm_size = "standard_b4ms"
    }

}
```

Note that you need to enable autoscaling as it is not a default setting for node pools. By setting the min and max count values, you are instructing Terraform to enable a minimum of one node and a maximum of six nodes in the default node pool. When this cluster is created, it will start with a single node in the node pool, despite what the node_ count setting is set to.

AKS allows you to define more than one node pool for your cluster, which can come in handy if you have workloads that are more resource intensive and can be isolated to different nodes to avoid taking resources from system workloads and processes. Next, we'll look at how to set up autoscaling for multiple node pools.

Tiered node pools with autoscaling

Having separate node pools for different workloads may seem like overkill at first glance. After all, if your application components are low-impact and run fairly lean, why even have an extra set of nodes spun up? In cases where your application needs to scale up certain operations while not impacting important system processes, the use of additional node pools can help balance that load but still give you the flexibility of autoscaling.

Following our example of using Terraform, we can add a resource that references the cluster but creates a new node pool for our application needs. The following snippet outlines how we can create a new node pool using zero nodes as the minimum:

```
resource azurerm_kubernetes_cluster_node_pool "app-pool" {
  enable_auto_scaling = true
  name = "appcompute"
  kubernetes_cluster_id = azurerm_kubernetes_cluster
    .aks-mtaeda-1.id
  vm_size = "standard_D8ads_v5"
  min_count = 0
  max_count = 10
  max_pods = 100
}
```

We are still enabling autoscaling, as well as setting the min and max node counts for the pool. In this case, you can see that the node SKU is a more powerful machine type than the SKU for the default pool. During the deployment process, you can set the nodeSelector property of the deployment to select this node pool instead of using the default pool.

Third-party plugins

As with any platform, native capabilities can generally be extended and ultimately lead to the creation of products that third-party providers offer to the community. There are a host of plugins for Kubernetes that are developed and maintained by individuals as well as companies, spanning a variety of use cases. Concerning scaling, we will be looking at one plugin that meshes well with our application's design paradigm – KEDA.

KEDA

KEDA, short for **Kubernetes Event-Driven Autoscaling**, is a plugin that enhances the autoscaling abilities of Kubernetes by allowing metrics to be set on pending events as opposed to sampled metrics for resources or deployments. That's not to say that events may not be tied back to some of those baseline metrics – but it does give you more options for how and when your application scales up (or down). There are over 50 scalers available to KEDA at the time of writing, but we will be focusing on two in particular: **Apache Kafka** and **Azure Application Insights**.

> **Important note**
> Please take a moment to review the links in the *Further reading* section of this chapter about the KEDA scalers mentioned previously.

The Apache Kafka scaler allows you to process triggers based on lag detected in processing messages for consumer groups and topics. This can result in an adjustment in the available number of replicas to reduce the lag detected. To take advantage of scaling using this mechanism, you will want to ensure the `scaleTargetRef` section contains a reference to the name of your Kafka deployment and at least the name of the consumer group associated. You may find that some groups don't require autoscaling to be configured since the volume of messages being consumed is relatively low. For other domains, such as **Equipment**, **Notification**, or **Identification**, a larger volume of messages may be consumed and autoscaling configurations for each consumer group would make more sense.

The Application Insights scaler allows you to process triggers based on specific metrics captured, whether those are out-of-the-box or custom metrics. If needed, further refinement of the metric can be done by adjusting the `metricFilter` property and including a clause that will target a specific outcome.

With that, we've seen how using various scalers for Kubernetes can allow you to control when and how many pods are allowed to scale when required. Next, we'll examine how to set up autoscaling for App Service.

Implementing autoscaling for Azure App Service

While a heavy focus on Kubernetes has been present throughout this book as well as this chapter, you may wish to leverage less complex **Platform-as-a-Service (PaaS)** components to host your application. Azure App Service offers a much lower barrier to entry than Kubernetes and allows you to craft different service types based on your needs. Autoscaling is one of the many features built into App Service and can be configured through different avenues. Two such avenues are those of Azure Monitor, the platform-level monitoring and alerting suite, and Application Insights. We'll also be examining some App Service specifics that can be applied to Web Apps, as well as API Apps.

Common platform options

One of the benefits of using cloud-native platform components is that you can apply them to just about any resource type within the cloud environment you're using. With Azure Monitor, you can tap into a large ecosystem of monitors, diagnostic information, and alerts to help you proactively and reactively manage your application's performance, availability, and reliability.

Azure Monitor

There are four main types of conditions that Azure Monitor uses to signal that a potential problem is afoot. These are as follows:

- **Metric alerts**: These are alerts based on a specific metric meeting or exceeding a predefined target value. These would include CPU usage, memory usage, and so on. An added dimension with metric alerts is time. You can set a threshold that invokes an alert should CPU usage hit 75% over a sustained 5-minute period.

- **Log alerts**: These are alerts that look for specific values or strings within Log Analytics that may be indicative of an error or issue.

- **Activity Log alerts**: These are alerts that specify when an action or event occurs that is not allowed or may indicate a compromised system. For example, the Activity Log captures information about creating and destroying cloud resources, as well as authentication auditing infrastructure-level failures.

- **Smart Detection alerts**: These are alerts that use the power of machine learning to adapt to your application's steady state, then use that information to detect anomalies and issue alerts based on those anomalies. Application Insights is one platform component that can utilize Smart Detection alerts.

While Azure Monitor is not solely about alerting, the ability to measure different platform aspects to determine if a resource needs to be scaled up can be helpful. Alerts can be auto-remediated using Action Groups, which are Azure constructs allowing for automation to fix issues that bubble up as alerts to the platform. These conditions also factor into the ability to scale App Services, more specifically App Service Plans, to match demand. We will look at how to configure autoscaling in the next two subsections.

Smart Detection alerts are particularly interesting because they allow us to leverage machine learning to detect failures and anomalies in our application with a simple configuration setting. That's not to say that other implications will not arise – cost, for example, may become a problem, especially if the volume of information that Application Insights is ingesting rises to hundreds of **megabytes** (**MB**) or even GB per day. Regardless of volume, we can still glean useful insights from Application Insights coupled with Smart Detection.

Application Insights

We have already seen how Application Insights offers a wealth of monitors, insights, and data points by simply creating an instance and integrating the SDK into our code base. The additional benefit of using Smart Detection is that it gives engineers a proactive diagnostic report when anomalies or failures are detected. While the use of Smart Detection does aid in offering automated alerts, setting up rules to scale other resources based on these alerts may not be practical.

One area of Smart Detection that does provide value is that of the **Settings** pane or the list of available Smart Detection alerts offered by Application Insights. These include dependency latency degradation, exception anomalies, failure anomalies, memory leak detection, response latency degradation, and trace latency degradation. The alert rules, conditions, and remediations for each of these items are pre-created when Smart Detection is enabled. This allows you to see how those alerts are constructed, and in turn, use the configurations to build custom autoscaling events in your App Services.

With more of an understanding of how integrated platform utilities such as Azure Monitor and Application Insights can provide us with important information relative to the performance of App Services, we can now turn to look at some special circumstances that are relative to two flavors of App Services: **Web Apps** and **API Apps**.

Adjustments for Web Apps

Web Apps are perhaps the most well-known implementation of App Services in Azure and have been around since the platform's inception. Many features and options have been added over the years – the ability to host and run a static web application, using

Docker containers as the mechanism to run your web application, along with a variety of application runtimes that can host applications. All App Services require an App Service Plan to be present, which controls the underlying infrastructure of the service as well as tie-ins to platform capabilities such as Azure Monitor.

To adjust the scaling settings for a Web App, or any App Service, you have two options. One is the ability to scale up, which is a manual operation and requires switching service plan types. The other is the ability to scale out, which can be done manually or automatically. This is where the logic used by Smart Detection alerts can come in handy, as it gives you an idea of how to construct custom scaling operations on an App Service Plan based on a variety of metric categories available to the plan. A full list of applicable metric types can be found in the *Further reading* section under *Azure App Services Web Site Monitoring*.

Common metrics such as CPU load, memory load, data I/O, and page load times are all available to you as means of setting up alerts that can drive autoscaling. Other more complicated metrics, such as lower-level TCP information about established and closing connections, can also be used.

Adjustments for API Apps

As API Apps used to be a separate type of App Service, it's common to see them referred to as such. When creating a new API App Service, you're creating a new Web App. Given the general construct that an API app is simply hosting a service or group of services that can be accessed via RESTful HTTP calls or path-based HTTP calls, there is little distinction between an API App and a Web App. Within the configuration of a Web App, there is a subsection that allows for configuring items such as API Management integration as well as service definition URLs that can load the service metadata from an OpenAPI or Swagger specification.

As for autoscaling, the same App Service Plan metrics and conditions for allowing an API App to scale out are the same as those for a Web App. Truly the only considerations you may look into would be whether to use API Management to act as a gateway to that Web App. API Management has a host of features and configuration options that can be beneficial but are beyond the scope of this book.

Having rounded out our look into Azure App Services and how they can be configured to allow autoscaling based on metrics, we can rest, knowing we have covered the main patterns of scalability as they relate to our application and our infrastructure.

Summary

In this chapter, we examined some key indicators of application performance, how easily resources can be adjusted to accommodate demands on the application, and how to set up automatic scaling for certain pieces of infrastructure. General scalability patterns commonly focus on being able to adjust the CPU or memory available to the application as a means to mitigate an increase in demand.

There are even ways of implementing new language features to help increase processing efficiency, as seen with the new `FileStream` object improvements in .NET 6. Through varying levels of complexity, we've looked at how to configure autoscaling for Kubernetes objects using metrics, thresholds, and even events. We've also looked at how to configure metrics, thresholds, and action groups to resolve issues with App Services. These patterns will become a vital part of your toolbox as you continue to architect event-based systems.

As we look forward to *Chapter 11, Minimizing Data Loss*, we will take what we have learned concerning scalability and pair it with reliability to suit the application and its data requirements.

Questions

1. Can network load force a scaling event for a virtual or physical network?

2. What performance indicators are most likely to be an engineer's primary focus when determining autoscaling patterns?

3. Are all the main performance indicators (such as CPU, network pressure, disk pressure, and memory) available to use as conditions for creating a horizontal pod autoscaler?

4. Instead of creating multiple node pools in an AKS cluster, wouldn't it be easier to increase the size of the nodes for the default pool?

5. If the Kubernetes best practice for creating metrics to use with HPAs is to not rely on external systems or objects, why would using a plugin such as KEDA be considered valid?

Further reading

*   *Design for scaling*, by Microsoft, available at `https://docs.microsoft.com/en-us/azure/architecture/framework/scalability/design-scale`

*   *File IO Improvements in dotnet 6*, by Microsoft, available at `https://devblogs.microsoft.com/dotnet/file-io-improvements-in-dotnet-6/`

- *Horizontal Pod Autoscale*, by Kubernetes, available at `https://kubernetes.io/docs/tasks/run-application/horizontal-pod-autoscale/`

- *Horizontal Pod Autoscale Walkthrough*, by Kubernetes, available at `https://kubernetes.io/docs/tasks/run-application/horizontal-pod-autoscale-walkthrough/`

- *Kubernetes Event-driven Autoscaling*, by Keda, available at `https://keda.sh/`

- *Kafka Scaler Documentation*, by Keda, available at `https://keda.sh/docs/2.8/scalers/apache-kafka/`

- *Application Insights Scaler Documentation*, by Keda, available at `https://keda.sh/docs/2.8/scalers/azure-app-insights/`

- *Azure Monitor Alert Types*, by Microsoft, available at `https://docs.microsoft.com/en-us/azure/azure-monitor/alerts/alerts-types`

- *Azure Monitor Alerts Best Practices*, by Microsoft, available at `https://docs.microsoft.com/en-us/azure/azure-monitor/best-practices-alerts`

- *Autoscale Best Practices*, by Microsoft, available at `https://docs.microsoft.com/en-us/Azure/azure-monitor/autoscale/autoscale-best-practices`

- *Action Groups*, by Microsoft, available at `https://docs.microsoft.com/en-us/azure/azure-monitor/alerts/action-groups`

- *Proactive Failure Diagnostics*, by Microsoft, available at `https://docs.microsoft.com/en-us/azure/azure-monitor/alerts/proactive-failure-diagnostics`

- *Azure App Services Web Site Monitoring*, by Microsoft, available at `https://docs.microsoft.com/en-us/Azure/app-service/web-sites-monitor`

11
Minimizing Data Loss

In this chapter, we will look at techniques for minimizing or eliminating data loss. Cloud-native development architectures bring an inherent risk of data loss through transient failures in separate services and/or connectivity between them. As such, we design applications for that failure, and event-driven architecture uses some key techniques to mitigate this risk of loss. We will explore paradigms, such as eventual consistency and guaranteed delivery, and learn how to contextually identify data that may be susceptible to loss, along with defining how much loss, if any, is acceptable.

In this chapter, we will cover the following key topics:

- Learn more about typical data consistency paradigms, such as immediate consistency and eventual consistency

- Learn how to identify and plan for potential data loss, including acceptable levels of loss and mitigation strategies

- For situations that require zero data loss, learn how to implement techniques that will guarantee the delivery of information to eliminate data loss

- Understand and account for implications of data loss, where data loss is assumed or expected

Technical requirements

For this chapter, you will require the GitHub source code found at `https://github.com/PacktPublishing/Implementing-Event-Driven-Microservices-Architecture-in-.NET-7/tree/main/chapter11`.

Preventing data loss

Ensuring the complete integrity of data for most monolithic applications is a static and manageable concept. Most underlying data stores are **atomic, consistent, isolated, and durable** (**ACID**). Once data has been received in the form of a request, a single compiled application will process it through to completion. Threats to the integrity of this operation are few and, more importantly, highly observable and consistent.

For example, a failure in the data store may consistently fail the whole application – or part of the application – until fixed. A failure of code presents itself as a bug that will fail the same way repeatedly for the same given request. Data corruption and loss in monoliths tend to be an all-or-nothing game. If data is being lost – you will know through *consistent failures to operate.*

Moving application designs to the cloud bring along two new challenges:

- Scalable architecture means multiple separate components are coordinated through many microtransactions to deliver an overall successful larger capability.

  - Given that microtransactions can occur at various times and take different durations to complete, we must understand the concept of **eventually consistent transactions** as an alternative to **ACID transactions**.

- The connection points between these components, and the component's availability, are subject to **transient failures**.

  - This could be caused by network failures or a platform service failure. While the cloud is designed to offer elevated levels of redundancy and availability to avoid these failures, they can and do still fail for a multitude of reasons. From an application perspective, it often appears random and inconsistent.

A dichotomy is created between these two new challenges.

On the one hand, we expect data to have occasional flaws in its integrity across components. Data should be **eventually consistent** across multiple services, *but not immediately.* As a result, we must design services to allow for a level of inconsistency over a brief period. On the other hand, many transient failures can happen across multiple components. This, in turn, means that there can be a loss of transactions between subcomponents that may go unseen.

So, at any given moment in time, when data is not consistent across components – are we waiting for that data to be eventually consistent – or are we experiencing data loss due to an unseen transient issue?

Data consistency paradigms

The two primary consistency paradigms to be aware of are referred to as ACID and BASE. We will define these acronyms in the following sections and understand what they are and how they differ. To help remember these acronyms, we can use an analogy from chemistry, where a *BASE* is the opposite of an *ACID*.

ACID paradigm/immediate consistency

As previously noted, **ACID** stands for **atomic**, **consistent**, **isolated**, and **durable**.

With immediate consistency, we get a guarantee that each transaction will be completed as a **single unit of operation**. Therefore, it will either entirely succeed or entirely fail. Take a hypothetical scenario of a consumer ordering a product to a new delivery address. In an immediate consistency model, the typical process for this transaction would be to first issue the address change and wait for positive confirmation that it was validated and written to the customer's profile. Next, the order would be created, at which point the new address is immediately available for read confirmation on the customer's profile. This is great for ensuring only accurate data exists at any given point in time. An order containing a customer address that is not on their profile is an error scenario that would fail validation and therefore fail to place the order.

BASE paradigm / eventual consistency

BASE stands for **basically available**, **soft-state**, and **eventually consistent**.

The *concept* of eventual consistency is simple. Mastering its *design and implementation* is a little more challenging but empowers a level of application scalability that is almost limitless.

Take any of the major social media platforms as an example. If every post ingested was expected to be written to a single data store, the bottlenecks would be quickly hit in all directions: memory, compute, bandwidth, disk I/O. The first line of defense for social media scalability is the partitioning of data – so when you post content, it won't necessarily be written to *all* global instances of the platform's data store; however, even if we just take a single region of the world, there is an incredible constant influx of new data that must be propagated around multiple instances of multiple services running in multiple locations. This propagation follows the concept of eventual consistency. Eventually, the data will be propagated; therefore, bringing consistency in data across the platform.

Eventual consistency is a great model for event-driven architectures. As events are written, they are, in effect, joining a queue with no strict governance on how that queue is processed. Subscribers to events may react immediately or at any point in the future. There may also be a delay depending on how many events have built up and what level of processing throughput the consumers can provide. So, working with the concept of data being eventually consistent is a natural fit for event-based systems.

Let us go back to the acronym of BASE:

- **Basically available**: This means that read and write operations are available but not guaranteed. This aligns well with transient faults that occur in a distributed cloud application. When interacting with a service that is basically available, there is no guarantee of consistency between the writes and reads.

- **Soft state**: This means that the state of the data could be fluctuating. As time passes, the expectation is that it will have a higher probability of converging (being consistent).

- **Eventually consistent**: This means that so long as the system keeps running, we can eventually expect any writes to be reflected in subsequent reads, and the state will be consistent.

Putting that theory to practice, we can consider payments for the **MTAEDA** service. The process of adding a new payment method to a customer's profile is an entirely different process from charging a customer when it comes to topping up their transit card balance. If a customer adds a new payment method and then wants to use that for a transaction in quick succession, there is no absolute guarantee that the validation and approval of the new payment method will have completed in time. Remember that this is by design. We have decoupled these processes to enable extremely high scalability and performance targets. Understanding this use case, the process for topping up a transit card balance needs to be designed around eventual consistency.

They are diverse ways you can build logic patterns to be tolerant of eventual consistency. For example, you could keep retrying a failing process to see if it will eventually work. It is cumbersome but much more commonly followed than you may think. A much better and well-known pattern is the **Saga pattern**.

The Saga pattern

Decoupled business transactions can be reformed into a flow, known as a **Saga**, such that they can eliminate the failure/retry pattern when dependent on eventually consistent data converging. They do this by changing the dependency order of transactions and using events to trigger the next transactions when consistency is achieved. In effect, this is like rebuilding a monolithic process, but by using events as the loose coupling, we avoid introducing hard dependencies or performance bottlenecks.

The example of adding and then using a new payment method is an example that can benefit from this pattern. A new payment method can be issued with a request increase the transit car balance. At this point, the top-up process can issue an event requesting the payment method be added to the customers' profile and leave the top-up request in a pending state. The payment profile service should (at its own time and pace) react to this event, validate, and add the new payment method. It will issue an event that the customers' payment method has been updated. The top-up process should (at its own time and pace) react to the event that a customer's payment method has been updated and look for any pending top-up activity that relies on that payment method.

While this may still sound a little cumbersome, it is important to notice how this decouples the process and allows a few *set-and-forget* processes that are coordinated through events rather than a single blocking process. Through chaining the transactions in this order (as a Saga), we see less polling and more event-driven activities. From this, we get eventual consistency and higher scalability.

Command query responsibility segregation (CQRS)

Eventual consistency demands quite different behaviors between transactions that modify the state of data and those that query it. The modifying state is authoritative and often results in a subsequent event confirming the modification and allowing the next steps in a flow to be triggered. The querying state must account for the fact that responses may be inconsistent and divergent from preceding modifications. This has driven the development of the **command query responsibility segregation (CQRS)** pattern in systems. It means that code should be written in such a way that commands (**modifications**, also often referred to as **mutations**) are segregated from queries (**reads**). This avoids writing code that may inadvertently go from one step to another with the natural expectation that external transactions are ACID in nature.

Having compared immediate and eventual consistency, your focus will likely be on embracing eventual consistency, Saga patterns, and CQRS. It is important to note that a well-functioning solution will use *both* ACID and BASE paradigms. We must focus on understanding the interplay between both paradigms in any given business transaction. Typically, you will see ACID transactions within a component and BASE transactions across components.

Identifying acceptable data loss

In designing systems that must deal with transient failures, data loss will inevitably occur. Many architectural patterns can be layered together to minimize the loss of data, but it is extremely challenging to guarantee that no loss will ever occur.

Later in this chapter, we will examine the implications of data loss and what we can do to compensate for it. Before getting to that, we must identify what acceptable and unacceptable data loss is. The benefit of this is narrowing the scope for unacceptable data loss, to which a high proportion of effort will go towards minimizing this loss.

Acceptable and unacceptable data loss

Defining what acceptable and unacceptable data loss is heavily influenced by the context of the action being performed and the business impact within that context. For example, if adding a customer's new payment method never succeeds and results in a data loss, it may be reasonable to assume the loss can be recovered by flagging the issue to the customer and asking for a repeat entry. However, if the **context** is within an e-commerce application, there may be significant reputational damage to consider when putting the responsibility on a customer to repeat such an action. Depending on *your business context*, you may consider the loss of a new payment method acceptable *or* unacceptable.

Another example of that may or may not be acceptable data loss is **navigation history**. Knowing where a customer has browsed on an e-commerce site is important for displaying recently visited products and other related products. However, if parts of this data are lost due to a transient failure, there is *no major functional failure* introduced into the system. Take another scenario where an application tracks the page a user is reading on an e-book. In this context, losing their latest page may mean resuming their reading from an incorrect position, either on the same device or a different device. In that context, it can be considered a *functional failure* of the application.

The critical functionality of your application will dictate what is or is not acceptable data loss.

Methods of data loss

Data can be lost in a system in many ways, from transient failures in connectivity and availability. We can abstract any **mechanism** of failure into two primary **modes** of failure:

- Failure to write an event
- Failure to consume an event

There is an assumption that the event broker is not going to fail internally. A key element of the event-driven architecture is that events can be handled consistently and reliably. What this means is if an event is received, it will be recorded by the broker. And if a request to consume events is made, it will be served. These are assumptions we can build high confidence in, as we will explore later.

However, all system components must internally function correctly and successfully connect with the event broker to create and consume the events. We can easily identify numerous methods of data loss:

- Network disconnection between components
- Network packet corruption
- Software bugs in a component
- Sudden termination of a running process
- Degradation/underperformance/timeout of a process

Many of these methods may sound like they are no different from a dedicated physical data center hosting a scaled application. However, on a cloud platform, there are many continuous reconfigurations of both physical and software-defined resources. Some of these are driven by maintenance activities, such as hardware replacement or operating system patching, while others are driven by operational desires, such as autoscaling or canary testing release strategies.

So, now we understand what acceptable data loss in our context is and the different methods that can introduce data loss, we can write code to accommodate for this. The failure to find a valid payment method can be written into the logic as a valid use case. In a monolithic application, it may be that this scenario can never occur! Designing systems that are susceptible to, and therefore consider these use cases, suggests that we are heading towards high scalability of overall system throughput.

Still, we made a big assumption that if an event is received by the broker, it will be served. Next, we need to understand the mechanisms that allow us to be so confident with this. Delivery guarantees are the way to do this.

Delivery guarantees

Event brokers typically provide several methods of delivery guarantees. Through the interfaces used to both produce and consume an event (API or SDK), the event broker can ensure how messages are delivered:

- At-most-once delivery

- At-least-once delivery

- Effectively once delivery

The delivery method we choose for any given event depends on which benefit we want to achieve. As we explore each delivery method, we will explore examples that highlight the benefits and trade-offs of each.

At-most-once delivery

At-most-once delivery means that the broker will ensure that a produced event is only delivered at most once (obvious, right?). What this means is as soon as the first consumer processes the event, the broker will not deliver the event to any other consumer. It is at-most-once delivery, so there is a possibility that an event will never be delivered:

- First, the producer does not wait for acknowledgment from the broker that an event is received. It will **fire and forget**. This is fantastic for performance and throughput. However, this also means the broker may never successfully receive the event. Yet the producer will just carry on as if it did.

- Second, a consumer that reads the event may crash or fail before it completes its processing. The broker is not aware of this failure and so will assume success and not serve the same event to any other consumer. Again, this is great news for performance, but hopefully, it is clear that this delivery guarantee is only suited for acceptable data loss in a system.

At-most-once code example – producer

We want to ensure that the **producer** is sending a message only once. If the producer crashes or the broker fails to receive the message, the producer should not retry sending the event. Either it *sends once* or *not at all*.

Looking at the `Send()` method, we can see this behavior is already in effect without making any more changes:

```
public async Task Send(string message, string
    correlationId, string causationId)
        {
            await Task.Run(async () =>
            {
                var messagePacket = new Message<int,
                  CloudEvent>()
                {
```

```
            Key = 1,
            Value = new CloudEvent()
            {
                Id = Guid.NewGuid().ToString(),
                OperationId = correlationId,
                OperationParentId = causationId,
                Message = message
            }
        };
        await _producer.ProduceAsync(Topic,
          messagePacket, CancellationToken.None);
        _logger.LogInformation("Producer sent the
        message: " + message + " - and with an Id
        of: " + messagePacket.Value.Id);
    });
}
```

The `ProduceAsync()` call might return an exception if the delivery report does not confirm a success. This would crash the thread. However, as the producer simply reacts to a real-time API request, the producer itself will not automatically retry sending the event again on restart. Either it *succeeded once* or it *failed*.

In a scenario where the event-producing code processes a stream or queue of incoming data, it is important to save read progress against this source prior to calling `ProduceAsync()`. This ensures that, after crashing and restarting, the producer will resume progress with the next message and not repeat an attempt to deliver the failed message.

At-most-once code example – consumer

As a precursor to illustrating how delivery guarantees are coded for the consumer, we need an operation that represents the **successful processing** of a consumed event. For simplicity, rather than coding a database write, we will instead simply write the event message body to a text file.

In the consumer's `MessageReceivedEventHandler` class, we will add a simple `File.WriteAllText()` call to do this:

```
public async Task Handle(ConsumeResult<int, CloudEvent>
    result, TelemetryClient telemetryClient)
```

```
        {
            await Task.Run(() =>
            {
                telemetryClient.Context.Operation.Id =
                    result.Message.Value.OperationId
                        .ToString();
                telemetryClient.Context.Operation.ParentId
                    = result.Message.Value.Id.ToString();
                telemetryClient.TrackTrace("Consumer
                    reacted to the message: " +
                        result.Message.Value.Message);
                File.WriteAllText(Path.GetTempFileName(),
                    result.Message.Value.Message);
            });
        }
```

Now that we have added this processing step, we can focus on achieving the at-most-once delivery guarantee configuration for the consumer.

The consumer code should process the message *after* it has acknowledged to the broker that the event has been consumed. This way, if the consumer happens to crash and restart while performing that processing, the broker will resume its progress with the *next* event and avoid resending the failed one again.

By default, the consumer configuration uses an underlying setting, known as **auto commit**, that controls how the consumer notifies the broker that it has successfully processed an event:

```
enable.auto.commit=true
```

With this setting, every time an event is processed, an **in-memory offset counter** is incremented to keep track of where the consumer is in the event stream. This is known as **storing the offset**. To maximize throughput performance, the value of this offset is only committed back to the broker every so often, as opposed to with each event processed. Although the time interval for the commit can be configured, it is always possible that the consumer could process events and locally update its offset but crash before this is committed back to the broker. That will not satisfy the at-most-once delivery guarantee. It means that there is a possibility of a consumer being sent the same event again after a crash and restart.

To achieve the desired guaranteed behavior, we must turn off the auto commit and replace it with a manual call to make the commit before we process the event.

In the consumer's `Main()` method, we simply set this configuration property to `false`:

```
var telemetryClient = scope.ServiceProvider
    .GetRequiredService<TelemetryClient>();

_config.EnableAutoCommit = false;

var service = new consumerService(_config, topicName,
    telemetryClient);
await service.Receive();
```

In the consumer background service, we call `Commit()` synchronously *before* actually processing the event locally:

```
protected override async Task ExecuteAsync
    (CancellationToken stoppingToken)
{
        while (!stoppingToken.IsCancellationRequested)
        {
                consumer.Subscribe(topicName);
                var result = consumer.Consume
                    (stoppingToken);
                consumer.Commit(result);
                await new MessageReceivedEventHandler()
                    .Handle(result, _telemetryClient);
        }
    }
```

Now we can be sure that the broker will only serve the consumer an event at most once. Should it crash while processing the message, the broker will resume serving events from the next offset in the stream.

These simple code changes require a lot of thoughtfulness to achieve the intended pattern. Now we can move on to at-least-once delivery and see how that changes the code in an adjacent way.

At-least-once delivery

At-least-once delivery means that the broker will ensure that a produced event is delivered at least once (still obvious, right?). What this means is all consumers can read the event, and the broker will continue to deliver it to any other consumer until such a time as a consumer acknowledges it has processed the event. It is at-least-once delivery, so there is a possibility that an event may be delivered more than once.

The producer must wait for acknowledgment from the broker that an event is received. If the broker does not acknowledge successful receipt, then the producer will repeatedly try until it does. This adds a throughput overhead, which will impact performance and scalability. However, this also ensures an event is received by the broker as is required for at-least-once delivery.

Any consumer that reads the event that crashes or fails before it completes its work will not be able to acknowledge successful delivery. Therefore, the broker will continue allowing consumers to read the event until it is acknowledged. Because of this, it is entirely possible that events are consumed more than once – but at least once. Overall, this is a burden on performance on both sides of the event, but hopefully, it is clear that this delivery guarantee is suited for unacceptable data loss in a system. At the same time, the consumers must account for the possibility of duplicate event processing.

At-least-once code example – producer

For the producer to ensure the event was received by the broker, it must examine the delivery state of the `ProduceAsync()` call. For completeness, it should also catch any exception that may occur and treat that as a failure.

We will wrap the `ProduceAsync()` call in a `do {…} while(...)` loop, and use the delivery result (or caught exception) to decide whether the delivery was successful or whether it needs to be attempted again:

```
bool brokerRecievedEvent;
            do
            {
                try
                {
                    var deliveryResult = await
                    _producer.ProduceAsync(Topic,
                      messagePacket,
                        CancellationToken.None);
                    brokerRecievedEvent =
```

```
                    deliveryResult.Status ==
                        PersistenceStatus.Persisted;
        }
        catch (ProduceException<int,
            CloudEvent>)
        {
            brokerRecievedEvent = false;
            Thread.Sleep(1000);
        }

    } while (!brokerRecievedEvent);
```

We are testing for a delivery result of `PersistenceStatus.Persisted`, which confirms acknowledgment from the broker that an event was received. There is another delivery result, `PersistenceStatus.PossiblyPersisted`, which indicates an event was transmitted, but there was no acknowledgment. In this scenario, it is possible the event was received by the broker but a network issue caused a failure to receive the acknowledgment. The event would be sent again by the producer so it can be certain of at-least-once delivery, even if this results in duplicate events.

With these changes, we can be sure that the producer will loop until a single event is successfully transmitted to the broker. In a more mature code base, you would not expect the producer to retry infinitely but instead track its own progress and store unacknowledged events so it is in a consistent state should there be a catastrophic system failure (such as a long-term network outage or unavailability of event brokers). The key goal to achieve is that an event must be sent at least once.

At-least-once code example – consumer

The changes required for the consumer are much simpler. All that is required is to perform `Commit()` back to the broker *after* the consumed event has been processed:

```
protected override async Task ExecuteAsync
    (CancellationToken stoppingToken)
    {
        while (!stoppingToken.IsCancellationRequested)
        {
            consumer.Subscribe(topicName);
            var result = consumer.Consume
                (stoppingToken);
```

```
          await new MessageReceivedEventHandler()
            .Handle(result, _telemetryClient);
          consumer.Commit(result);
    }
  }
```

We know that the local offset will automatically update in memory as the consumer receives new events. However, by only triggering the synchronous commit *after* processing the event, we can be sure that any crash during that processing will result in the same event being served up to the consumer again on restart.

The consumer must also handle the possibility of duplicate events. For example, the broker could receive an event from a producer but fail to have its acknowledgment received. In the same way, so could a consumer process an event but fail to commit its progress back to the broker. For this to be acceptable, the consumer's process should be **idempotent**.

For our simplified `File.WriteAllText()` approach, we just need to use a predictable filename based upon the event being received to make it idempotent:

```
telemetryClient.TrackTrace("Consumer reacted to the
  message: " + result.Message.Value.Message);
              var filename = Path.Combine
                (Path.GetTempPath(), result
                  .TopicPartitionOffset.ToString());
            if(!File.Exists(filename))
                File.WriteAllText(filename,
                  result.Message.Value.Message);
```

Now if the same event is sent to the consumer multiple times, there is no detrimental impact on the overall process it serves.

With the code changes in this section, you can see how the at-least-once delivery guarantee adds some additional overhead for both the producer and consumer. This is a performance trade-off as delivery guarantees become more reliable and observable. Next, we will tackle the most reliable but least performant delivery guarantee available.

Effectively-once delivery

Effectively-once delivery means that the broker will ensure that a produced event is delivered *effectively* once (too obvious, right?). What this means is only one consumer can read the event, and the broker expects acknowledgment it has processed the event.

This delivery guarantee is also known as *exactly* once. There are many reasons why the use of the term *exactly* is hotly debated (see the *Further reading* section). For the purposes of this book, we will continue calling it *effectively* once and avoid those arguments.

One approach to achieving effectively-once delivery requires transactional scopes on both the producer and the broker. As these are separate processes, there is always a risk that one scope will successfully commit and the other will fail before completing its commit. We can keep this risk extremely low by keeping the scope committals side by side in code without any other complexities between, but this dual transaction scope synchronization simply can not guarantee 100% reliability (although it won't be far off!).

Just like at-least-once delivery, the producer must wait for acknowledgment from the broker that an event was received. If the broker does not acknowledge successful receipt, then the producer will repeatedly try until it does.

Any consumer that reads the event must respond with an acknowledgment. Otherwise, the broker will allow another instance (or restart) of the consumer to read the event again, and it will continue to do this until exactly one delivery is confirmed.

Effectively-once delivery has the greatest performance trade-off of all three delivery methods. This guarantee is best suited for data that is considered an unacceptable loss in a system where accounting for the possibility of duplicate events is not feasible or desirable (that is, the consumer will not handle duplicate events with **idempotency**).

Effectively once code example – producer

To achieve effectively-once delivery from the producer, we must again ensure that the producer receives an acknowledgment from the broker. But, in the scenario where there is a failure in receiving that acknowledgment, any retry by the producer must not result in a duplicate event in the broker. Additionally, we must ensure that any saved progress against the producer's source is atomically tied to a successful commit with the broker.

To tackle the first challenge, we can enable a configuration setting in the producer client, `enable.idempotence`. This is disabled by default. When enabled, each producer requests a unique producer ID from the broker. Each event sent to the broker will now include that unique producer ID. It will also include an auto-generated sequence number. So now, when the producer sends a retry event, the broker has a way to identify it, effectively ignore it, and simply send back another acknowledgment.

To achieve this, all we need to do is enable idempotence in the producer client:

```
var producerConfig = new ProducerConfig();
producerConfig.EnableIdempotence = true;
```

We have removed the creation of duplicate events from retires, meaning we have *exactly-once* delivery for a single instance of a producer. *But* what if the producer crashes before receiving the acknowledgment? A new instance of the producer will be given an entirely new producer ID by the broker. It will pick up from its **internal saved progress marker**, which will result in either duplicate events or lost events depending on whether progress is saved before or after the event is produced (**at least once** *or* **at most once**).

The only way around this is to atomically tie together saving local progress in the producer with the acknowledgment that the event was successfully received by the broker.

We can use `TransactionScope` when saving local progress so that the progress save point is rolled back should there be a failure within the scope. Kafka also offers a transaction scope so it can roll back any received event if the broker scope does not complete successfully:

```
do
      {
            try
            {
                  using (var localTransaction = new
                    TransactionScope())
                  {
                        _producer.BeginTransaction();
                        var deliveryResult = await
                        _producer.ProduceAsync(Topic,
                        messagePacket,
                        CancellationToken.None);
                        brokerRecievedEvent =
                        deliveryResult.Status ==
                        PersistenceStatus.Persisted;
                        _logger.LogInformation
                        ("Producer sent the message: "
                        + message + " - and with an Id
                        of: " + messagePacket
                        .Value.Id);
                        //Save internal progress here
                        if (brokerRecievedEvent)
                        {
      // Only risk of uncoordinated failure is a crash
```

```
        between these two lines...
                        _producer.CommitTransaction();
                        localTransaction.Complete();
                    }
                    else
                    {
                        _producer.AbortTransaction();
                    }
                }
            }
            catch (TransactionAbortedException)
            {
                _producer.AbortTransaction();
                brokerRecievedEvent = false;
            }
        } while (!brokerRecievedEvent);
```

There is an intertwining of producer local and broker remote transaction scopes that ensures synchronization between the successful saving of progress and the successful receipt of events with extremely high (but not flawless) reliability.

There is a way to eliminate that small risk of failure between the commitment of local and broker transaction scopes, and that is to also use Kafka as the state storage for our local progress. In that scenario, we only have one transaction scope, which lives entirely on the broker. This can come with its own niche challenges depending on the nature of the producer application, which is beyond the scope of this book.

Effectively-once code example – consumer

Achieving effectively-once delivery to the consumer follows a similar pattern to the producer. However, instead of using a transaction scope on the broker, we can instead control the internal offset values and commit calls to achieve the same effect.

To begin with, we must turn off the auto increment of the internal offset counter, so we can manually control this as part of the transaction scope:

```
var telemetryClient = scope.ServiceProvider
    .GetRequiredService<TelemetryClient>();



        _config.EnableAutoCommit = false;
```

```
_config.EnableAutoOffsetStore = false;

var service = new consumerService(_config,
    topicName, telemetryClient);
```

Now setting the current offset and committing this back to the broker must be done manually in the code.

Let us look at how this is used with the transaction scope when consuming an event:

```
var result = consumer.Consume(stoppingToken);
            Offset originalOffset = consumer.Position
                (result.TopicPartition);
            try
            {
                using (var localTransaction = new
                    TransactionScope())
                {
                    consumer.StoreOffset(result);
        //Save internal progress here
                    await new MessageReceived
                    EventHandler().Handle(result,
                    _telemetryClient);

    // Only risk of uncoordinated failure is a crash between
        these two lines...
                    consumer.Commit(result);
                    localTransaction.Complete();
                }
            }
            catch (TransactionAbortedException)
            {
                //Restore the offset position
                consumer.StoreOffset(new
                TopicPartitionOffset(result.
                TopicPartition, originalOffset));
        consumer.Commit(result);
            }
```

Before processing the event, we capture the current offset position. We can then increment the offset to the current event being processed. After the event is processed, we commit the offset position to the broker and the local transaction scope, which will contain any saved progress.

Should the event processing fail and the local transaction is aborted, we will reset the offset to the previously saved value.

Again, we could save local consumer progress to Kafka. In that scenario, the consumer would also be producing its own events to track that progress. The producing service provides access to the broker transaction scopes.

We have covered the delivery guarantees in some practical detail. But how do they relate to the MTAEDA application? And what are the implications of each?

Data loss implications

There are many scenarios we could evaluate to find the optimal delivery guarantee for a given process. As an example, each time a mechanical barrier counter increments, this is sent via an event to the maintenance tracking service. If each event simply contains a barrier ID, then we have the decision to make in the accuracy of the counter driving the maintenance schedule.

At-most-once delivery will ensure we do not overcount the use of each barrier due to system failures that would otherwise result in duplicate events. However, it also means we might undercount for the same reasons. This could result in late maintenance, meaning a higher probability of barrier failure.

At-least-once delivery will ensure we do not miss any counts in the use of a barrier. However, it also means we might overcount through duplicate (retry) events. This could result in early maintenance, meaning a higher cost to operate. It will also impact system scalability and throughput.

Effectively-once delivery would ensure that the barrier usage count is accurate. However, how does the even greater performance impact of this guarantee measure up against the mistiming of maintenance?

As we have previously discussed, this design decision (and the many other event scenarios that require similar consideration) are deeply rooted in the real-life domain that our event-driven architecture supports.

Summary

In this chapter, we looked at the potential for data loss in a distributed system and examined how we can use event delivery guarantee patterns to appropriately sacrifice performance for a lower loss risk.

We learned the differences between ACID and BASE paradigms and how eventual consistency is a key consideration of distributed event-driven data architecture. After understanding how to identify and evaluate acceptable data loss, we looked at how to apply different delivery guarantees between our services and the event broker to match the need. Finally, we considered what the impact of data loss might be and how the context of the real-life domain challenge we are addressing influences our design response to that potential loss.

In the next chapter, we will switch focus to service and application resiliency.

Questions

1. Which data consistency paradigm ensures a consistent state at any given moment?

2. Which data consistency paradigm cannot ensure a consistent state at any given moment?

3. What development pattern is used to avoid assumptions of a consistent state from a data paradigm?

4. What data consistency paradigm can support hyper scalability?

5. If an event-driven transaction can accept data loss but is not idempotent, what is the most suitable delivery guarantee configuration?

6. If an event-driven transaction cannot tolerate data loss and is idempotent, what is the most suitable delivery guarantee configuration?

7. If an event-driven transaction cannot tolerate data loss and is not idempotent, what is the most suitable delivery guarantee configuration?

Further reading

- *Processing guarantees in Kafka* – by Andy Bryant, available at `https://medium.com/@andy.bryant/processing-guarantees-in-kafka-12dd2e30be0e`

- *Exactly-once Support in Apache Kafka* – by Jay Kreps, available at `https://medium.com/@jaykreps/exactly-once-support-in-apache-kafka-55e1fdd0a35f`

- *Kafka Transactions and Exactly-Once Processing* – by SmallRye, available at `https://smallrye.io/smallrye-reactive-messaging/3.18.0/kafka/transactions/`

- *Saga distributed transactions pattern* – by Microsoft, available at `https://learn.microsoft.com/en-us/azure/architecture/reference-architectures/saga/saga`

- *CQRS pattern* – by Microsoft Open-Source Community, available at `https://learn.microsoft.com/en-us/azure/architecture/patterns/cqrs`

- *Data Consistency In Microservices Architecture* – by Dilfuruz Kizilpinar, available at `https://resources.experfy.com/bigdata-cloud/data-consistency-in-microservices-architecture/`

12
Service and Application Resiliency

In *Chapter 10*, *Modern Design Patterns for Scalability*, we reviewed some patterns for addressing scalability within our application. An equally important concept is that of resiliency, meaning the likelihood of a service or application handling errors or exceptions at runtime. Accounting for changes in traffic and usage for components does not mean much if the components themselves do not function properly, or do not provide any clear means of tracking down the root causes of issues.

There is always the possibility that a cloud service could experience a brief outage or an extended period of downtime that is completely out of the hands of those who maintain the application or platform.

In this chapter, we'll cover the following:

- Resiliency through cloud-native patterns
- Redundancy and enabling business continuity
- Graceful communication

By the end of this chapter, you will be able to review and implement cloud-native software patterns that will improve the resiliency of your application. We'll also examine redundancy techniques for ensuring access to data in the event of a service outage locally, regionally, or platform-wide, and learn how to craft user-friendly messaging that clearly communicates the issue at hand but does not cause the application to completely fail.

Technical requirements

You will find all the code examples for this chapter available in the folder for this chapter on GitHub at `https://github.com/PacktPublishing/Implementing-Event-Driven-Microservices-Architecture-in-.NET-7/tree/main/chapter12` and `https://github.com/PacktPublishing/Implementing-Event-Driven-Microservices-Architecture-in-.NET-7/tree/main/src`.

> **Important note**
> The links to all the white papers and other sources mentioned in the chapter are provided in the *Further reading* section toward the end of the chapter.

Resiliency through cloud-native patterns

Application resiliency can be measured by how durable and recovery prone a service or application is. There are different paradigms for analyzing and implementing resiliency in applications, including infrastructure and software design patterns. For the purposes of this chapter, we will be examining software patterns that leverage cloud-first architectural patterns to bolster application resiliency.

While we can review the architectural patterns using the Azure architecture center as well as various sources for software resiliency, we're going to be looking at a library that's fairly prevalent, especially in cloud-first development circles. This library is called **Polly.net**. Polly.net is a library that allows you to implement several different types of policies in either a synchronous or asynchronous manner to address specific issues or to combat known problems with cloud service transiency, as well as advanced error handling techniques, such as the **circuit breaker pattern**.

To familiarize ourselves with Polly.net, let's look at a simple example where we can implement both a circuit breaker policy and a retry policy. Retry policies are one of the most common ways in which to establish resiliency, as you can determine actions to take after the retry count has been exceeded. As well as the retry and circuit breaker patterns, there are several other patterns that you can implement using Polly.net. These include the following:

- **Bulkhead isolation**: This pattern purposely restricts the number of resources available to a caller, as well as the number of requests that can be processed.

- **Cache**: Using this policy will force the method being executed to pull from the cache first. If the value doesn't exist, it will fetch the value from the data store and put it into the cache.

- **Timeout**: This guarantees a maximum time to wait. For example, a timeout policy set with a limit of 5 seconds will fail if the operation takes longer than 5 seconds.

- **Fallback**: This provides a way to return an alternative value for a method when the method call fails.

- **Rate limit**: Sometimes seen as throttling, this limits the number of consecutive calls from a caller if the number reaches a certain threshold.

- **PolicyWrap**: This is a lightweight wrapper that allows you to combine multiple policies into one and execute them as a part of a group.

It's possible to group multiple policy implementations together using the `PolicyWrap` object. For example, you could group retry and fallback policies together to help a method retry its execution and return a default message upon failure. The fallback policy is somewhat like the circuit breaker in that an alternative message or function can be used to communicate success or failure. They differ, however, in that fallback policies could be executed many times and not halt traffic to a particular method call, whereas circuit breakers will halt traffic for a set period.

For the purposes of our example, we're going to be using a simple example to make sure that we're assigning a retry policy to a specific block of code. Once we have successfully added that code and tested it, we will move on to implementing a similar strategy in the sample application itself:

1. To get started, create a new project in Visual Studio. You can use a console app template, a Web API or REST API template, or any other template of your choosing. Right-click on the project and select **Manage Packages** from the context menu. We will be installing Polly from NuGet.

2. Next, we will add a new class to the project that we have just created to house the logic we want to execute for the service. You may call this class anything you like; however, for this example, we will name it `ServiceBusinessLogic`.

3. In the new class we have just created, let's add a private variable of the `AsyncPolicy` type to hold a reference to our retry policy. In the class constructor, we can instantiate the `AsyncPolicy` object as such:

```
retryPolicy = Policy.Handle<Exception>()
  .WaitAndRetryAsync(3, retryAttempt =>
        {
            return TimeSpan.FromSeconds
               (retryAttempt++);
        }, (e, t) =>
        {
            Console.WriteLine($"Exception caught
               for retry: {e.Message}");
            // Add code to perform additional
               operations if needed
        });
```

4. This will use the `Policy` static object to assign a handler for exceptions (in this
    case, any exception), as well as set the number of retries and provide a function that
    will return an incremented `TimeSpan` object for the delay between retries. For this
    example, we are starting with 1 second and simply incrementing by 1 after each
    retry attempt. In some scenarios, a more gradual backoff timing may be desired, in
    which case, using something like `Math.Pow(2, retryAttempt)` may be more
    useful.

5. Now, add a method that is publicly accessible with a return type of `string` and
    name it `TestMethod`. Within the body of this method, we need to do two things:
    invoke the policy and populate the method with some functionality to return
    a value. Let's first start by creating the core functionality of the method:

```
return await Task.Run(async () =>
            {
                try
                {
    // Run some nonsense code here to get a failure
                        var rdm = new Random();
                        if (rdm.NextInt64() % 2 == 0)
                        {
                            throw new
                            InvalidOperationException
                               ("Random exception");
                        }
```

```
                    if (string.IsNullOrEmpty
              (result.Trim())) throw new
              ArgumentNullException(paramName:
              nameof(result), message: "You
              forgot to add a value.");
                    _returnMessage = $"String
                    values can be anything you
                    like: {result}";

                    return PolicyResult<string>
                    .Successful(_returnMessage,
                    null).Result;
              }
              catch (TimeoutException ex)
              {
                    Console.WriteLine("Handled");
                    return PolicyResult<string>
                    .Failure(ex, ExceptionType
                    .HandledByThisPolicy, null)
                    .Result;
              }
        });
```

This code is doing two things. First, it is generating exceptions at random to help trigger the retry policy at different intervals. Second, it is throwing an exception when a blank value is passed into the method. During testing, experiment with different values, including values that are just a combination of spaces. You will see messages from the policy as it retries the invocation due to the exception.

6. With the functionality implemented, we will next look to invoke the policy as a part of the method itself. The following example shows how to invoke the ExecuteAsync method of the policy to allow it to monitor the code's execution. If any timeout exceptions are caught, the retry pattern will automatically kick in to handle retrying the call:

```
        public async Task<string> TestMethod(string
            result)
        {
            return await retryPolicy.ExecuteAsync
```

```
<string>(async () =>
{
    return await Task.Run(async () =>
    {
        ... (code goes here)

    });

});

}
```

7. This gives the method additional cushioning to keep trying to execute in the event of a transient issue. Let's now turn to the `Program.cs` file and add a route to the API to test out our policies. Toward the bottom of the file, insert the following:

```
app.MapGet"/testbreaker", async
    (string? stringValue) =>
{
    return await breaker.TestMethod(stringValue ??
        "Nothing");
});
```

8. Debug the code using Visual Studio by either using the context menus to start debugging or by hitting the hotkey associated with running the project in debug mode (normally *F5*). Once the Swagger page loads, expand the `testbreaker` operation, as shown in *Figure 12.1*:

Figure 12.1 – Expanded operation in Swagger UI

9. Next, click on **Try it out** and enter any characters you like in the textbox. Once finished, click the **Execute** button. If the operation completes successfully, you should see something similar to *Figure 12.2*:

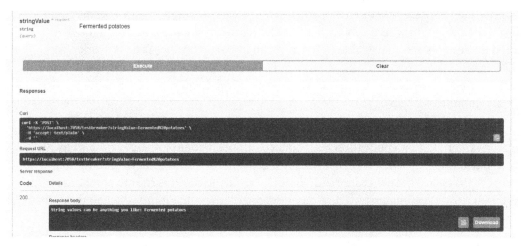

Figure 12.2 – Successful execution of the TestMethod operation

10. Continue experimenting with different values, including passing in spaces. Pay attention to the program output window in Visual Studio, as you should start to see messages appearing there stating that a retry is occurring. When that happens, you should see text in that window that is similar to *Figure 12.3*:

Figure 12.3 – Program output showing retry execution in progress

Having tested out an isolated example of using Polly to implement a retry policy, we will now turn our attention to implementing policies in our sample application. Using retry and circuit breaker policies can help us not only to increase resiliency but also to leverage redundancy to help better preserve operations from failing.

Redundancy and enabling business continuity

Whether an outage occurs due to a transient error or a regional failure, ensuring continuity can take different forms. For highly critical applications, data, and files, having a good business continuity plan in place will help business functions to keep carrying forward. Normally, business continuity is managed as a separate function but ultimately contributes to a larger disaster recovery plan for the organization.

A common scenario where having a business continuity plan helps is when there is an outage related to office documents or functionality. Many companies have a shared or distributed filesystem where spreadsheets and documents are often stored, and when shared documents are unavailable to users, it can impact a business's daily work effort. While many online services, such as Google Docs and Microsoft 365, can offer a much more durable means of managing and accessing critical documents, even those services can fall victim to outages.

Another common scenario is that of an internal application that is used by a group of employees: should that application get knocked offline, it could impact productivity. If that application is built in a way that does not account for application data redundancy (and perhaps even infrastructure redundancy), an outage could turn into something much more impactful. Keeping factors such as unexpected downtime in mind when designing an application is something that will serve you well once the application is live.

Event retries and continuity

Since there are no examples of either scenario that would come into play with our application, we instead look to areas where redundancy would benefit us. In the last section, we looked at software patterns that can help manage resiliency aspects as well as perform different actions based on the success or failure of those policies. Next, we can examine how to augment the resiliency policies with redundancy in the equipment service:

1. First, let's open the equipment domain solution in Visual Studio. The area of interest we will be taking on is the publisher responsible for logging turnstile usage. Navigate to the `Command API` project and open the `TurnstileController` file.

2. You will see two methods at the bottom of the class addressing the passenger ingress and egress events for the domain. Here is where the policy calls will be implemented. We do need to create the retry policy first, so we will be adding the following code to the `TurnstileController` constructor:

```
asyncPolicy = Policy.Handle<Exception>()
  .WaitAndRetryAsync(3, retryCount => {
    return System.TimeSpan.FromSeconds(retryCount++);
});
```

3. This will perform a simple incrementation of the waiting period with a maximum of 3 seconds. In order to apply the policy to the ingress and egress, we will add the following code to each of the methods:

```
return await asyncPolicy.ExecuteAsync(async () =>
{
    return await Task.Run(async () =>
    {
        return await RaiseEvent(PassengerIngressEvent
            .Create(turnstileId));
    });

});
```

4. We will want to ensure that the code within each method raises the right event based on the method signature. The preceding code creates a new `PassengerIngressEvent` event, and the code used for the egress event should create a new `PassengerEgressEvent` event.

Local data redundancy

With this added, we can now turn to the redundancy needs. Aside from having multiple copies of the service running, we also need to make sure that, in the event of a communication failure, the event data is persisted locally. This can be performed in several different ways. We will examine writing the event data out to a server at the station, which would be running a separate service to specifically handle a scenario, such as a turnstile not being able to communicate with the outside world. We are assuming that the networking within a station is hardwired and that each turnstile has a direct line.

The new service will simply run as a background process, listening for any calls to save event data. It will also periodically poll for connectivity to the IoT Hub, which serves as the gateway between stations, turnstiles, and any Azure-based components. Once connectivity is established, the service will attempt to republish the events that were not originally sent to the hub.

Since we don't want to inject any additional information into the events that are written to the station's server, we will instead write logic into the service to allow the events to be saved to paths on the machine that correlate to the event type by using the following steps. This can serve two purposes: to allow keeping event data grouped together logically and to facilitate republishing the events to the appropriate topics in Kafka.

1. Let's start by opening the MTAEDA.Core solution in the src folder. We will be adding a new event and event handler to the Core project to allow us to raise a retry failback event and, subsequently, handle that event in the new listener project we will create.

2. Create a new class called RetryFailbackEvent, which implements IDomain Event. Then, create a new class called RetryFailbackEventHandler, which implements IDomainEventHandler<RetryFailbackEvent>.

3. Then, create new folders in the project for Events and EventHandlers, placing the new classes in their respective folders. Your project layout should look like *Figure 12.4*:

Figure 12.4 – Expanded project view in MTAEDA.Core

4. For the implementation of RetryFailbackEvent, we will have the DomainData property implemented, but will also want to add a few more properties to capture important information, as shown in *Table 12.1*. These properties will all be set by passing in values to the static Create method. Please refer to the source code on GitHub for more information about that method implementation.

Property	Type	Description
Action	String	The action being performed – likely the calling method's name
Domain	String	Originating domain that is sending the event information
FailoverSite	String	Address to send the event information to
TransactionDate	DateTime	Timestamp for the event

Table 12.1 – Properties for RetryFailbackEvent

5. Next, we will look at the implementation of the IDomainEventHandler interface in the RetryFailbackEventHandler class. In the body of the Handle method, we want to take the information that has been forwarded in the event and persist it to disk. Write the following code inside of the Handle method to persist the failback event data:

```
public Task Handle(RetryFailbackEvent
    notification, CancellationToken
    cancellationToken)
{
    return Task.Run(() =>
    {
        File.WriteAllText($"{Environment
        .ProcessPath}\\{notification.Domain}
        \\{notification.Action}
        \\{notification.DomainData?
        .Cast<CloudEvent>().Single().Id}
        .json",
            notification.DomainData);
    });
}
```

6. Using a standard expression, we're able to write out the event's contents to a JSON file following a known file path. It's likely that storing the files in the same directory as the running service may be an area you wish to modify, especially if you're able to utilize a data or non-OS drive for file storage.

7. The next part of the solution is going to be the project containing the listener and the service endpoint that will allow the turnstiles to pass event information along if a communication error occurs. We will add a new ASP.NET Web API project to the

Listeners folder and name it `MTAEDA.Equipment.RetryFailback`. Be sure to leave the **Use controllers** checkbox unchecked, as we will be using a minimal API for this particular service, as shown in *Figure 12.5*:

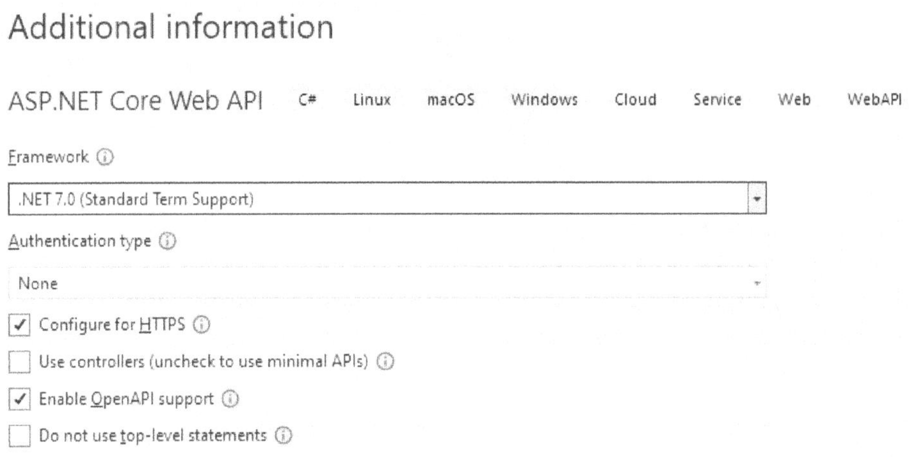

Additional information

ASP.NET Core Web API	C#	Linux	macOS	Windows	Cloud	Service	Web	WebAPI

Framework ⓘ

.NET 7.0 (Standard Term Support)   ▾

Authentication type ⓘ

None   ▾

☑ Configure for HTTPS ⓘ

☐ Use controllers (uncheck to use minimal APIs) ⓘ

☑ Enable OpenAPI support ⓘ

☐ Do not use top-level statements ⓘ

Figure 12.5 – Additional information in the new project wizard

8. Remove the boilerplate `WeatherService` code from the project and, instead, map a new `POST` method to allow event information to be passed from the turnstile:

```
app.MapPost("/failback", (context) =>
{
    return Task.Run(() =>
    {
        HttpStatusCode returnCode = HttpStatusCode.OK;
        HttpContent returnContent;

        RetryFailbackEvent evt =
          RetryFailbackEvent.Create("http://
            failover.site", "equipment",
              "turnstileIncrement");
        using (StreamReader sr = new StreamReader
            (context.Request.Body))
        {
            try
            {
                evt = JsonConvert.DeserializeObject
```

```
                        <RetryFailbackEvent>(sr.ReadToEnd())
                            ?? evt;
                    RetryFailbackEventHandler handler =
                        new RetryFailbackEventHandler();
                    handler.Handle(evt,
                        CancellationToken.None);
                    returnContent = JsonContent.Create(new
                        { message = "Operation Successful"
                            });
                }
                catch (Exception ex)
                {
                    returnContent = JsonContent
                        .Create(ex);
                }

            }
            return new HttpResponseMessage(returnCode) {
                Content = returnContent };
        });
    });
```

9. There is one last piece we must create for this service – a background worker that will poll the IoT Hub to determine whether it is reachable or not. Writing out events that have exceeded a retry threshold is great for retaining information but does not ensure the data makes its way to where it needs to be. Create a new class named IotHubConnectivityWorker and ensure it derives from the BackgroundWorker class. There will be two key elements of this class: the loop that determines whether connectivity to the hub exists, and the use of an IEventWriterProvider interface to ensure the saved events are processed appropriately.

10. The main functionality of the IotHubConnectivityWorker class is found in the DoWork method, which we override to facilitate customization. The following code can be placed into a try-catch block to start moving through any files that were written out as a result of RetryFailbackEvent being handled:

```
var baseDirectory = e.Argument?.ToString() ??
    Environment.CurrentDirectory;
```

```
                    if(!Directory.Exists
                    (baseDirectory)) throw new
                    IOException($"Directory does not
                     exist: {baseDirectory}");


                    foreach(var dir in Directory
                    .GetDirectories(baseDirectory))
                    {
                   // Each directory will be a domain
                        var domainName =
                          Path.GetDirectoryName(dir);
                        foreach(var action in
                          Directory.GetDirectories
                            (dir)) {
                  // Each directory will be an action
                            var actionName =
                            Path.GetDirectoryName
                            (action);
                            foreach(var file in
                            Directory.GetFiles(action,
                            "*.json"))
                            {
        // Attempt to deserialize each JSON file and send
          it along
                                var evt = JsonConvert
                                .DeserializeObject
                                <CloudEvent>(file);
                                evt.Type = actionName;
                                evt.Source = new
                                Uri($"http://
                                {Environment
                                .MachineName}
                                .RetryFailbackService/
                                {domainName}/
                                   {actionName}");
                                _producer.Send(evt);
```

```
            }
        }
    }
```

11. For the catch block, it's important to capture the exception but not to break the loop execution, since it does need to continue running should more events be sent to the service. We will use the logging service injected via the `ILogger` interface to record the error and continue execution, as we are expecting to encounter exceptions during this process.

> **Important note**
>
> In the `Infrastructure` project of the equipment domain, we'll be adding a new implementation of `IEventWriterProvider` to allow for sending events to an Event Hub endpoint, which is a common implementation when working with the IoT Hub.

Infrastructure continuity

Our focus can now turn to the Azure resources that need to be in place to help facilitate event publishing. *Figure 12.6* provides an overview of the path an event would take if it were to be sent to the `RetryFailback` service and then forwarded once connectivity to the hub was restored:

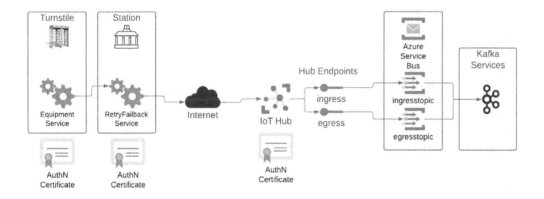

Figure 12.6 – Diagram of an end-to-end solution for equipment event redundancy

Let's walk through a few of the areas in the diagram in a bit more detail. We are already familiar with the Azure IoT Hub and the logical connection between it and other services. Certificate authentication between devices and the IoT Hub is a common and supported pattern, and extending the certificate to the `RetryFailback` service allows us to maintain the same authentication method and ensure communications will be permitted. There are two endpoint configurations that are key to helping route turnstile events once the IoT Hub has received them. These endpoints are set up to forward specific events to topics located within an Azure Service Bus instance.

Service Bus has been selected over Storage Containers, CosmosDB, and Event Hubs in this instance for a couple of reasons. The cost per transaction is typically lower with Service Bus than with the other endpoint types. Also, using Service Bus allows for the potential of integrating other systems with our sample application without having to retrofit connectivity to Event Hubs.

Finally, we can configure Kafka, as we did in *Chapter 3*, *Message Brokers*, to listen for events that get written to the two topics in the Service Bus instance. This allows for seamless integration into Kafka and allows further event handlers to fire, depending on the needs of the domain.

Now that we have addressed a redundancy need with event data, we can focus on ensuring downstream consumers can make sense of what has happened. With the graceful degradation of services comes the need for succinct and graceful communication.

Graceful communication between services

Everyone at some point has experienced an issue with a website or application where an error message will pop up that means absolutely nothing to the end user. Sometimes, it's no error message at all. Other times, it can be a complete stack trace dump that appears in the error message. Either of these scenarios is far from optimal, as the impact on the user is detrimental and could potentially share information about the code base if messages aren't adjusted.

The same can be true of messages tracked in logs. If no information is sent about a potential error or exception, finding out what happened – and how to fix it – becomes very challenging. It's important not only to the usability of the system but the maintenance of it as well to have clear, meaningful messages meant for each type of audience.

A common way to manage exception messaging is through the development of exception types that derive from `System.Exception`. For example, the `NullReferenceException`, `FileNotFoundException`, and `InvalidOperationException` types are all derived from the base `Exception` class. A common library of `Exception` types could be made available to all development teams using a shared library, or each team could create and maintain those custom types.

In most cases, however, this still may not be the preferred approach to communicating with end users. Often, microservice developers will make use of custom HTTP return codes to denote specific results back to the user. Whether that return code is caught on the frontend and rendered to the user with a client-side library or caught on the backend and translated into a meaningful message, the onus is on the developer to ensure a human-friendly way of conveying an issue with the application.

Common results

Let's revisit the `BaseCommandController` file and jump back into the `RaiseEvent` method. We've already seen that two error codes are being accounted for during the method's execution. As many of the domains have APIs that work with HTTP return codes, it would make sense to have a common error-handling framework that gives all API projects in each domain a consistent means of returning status messages based on HTTP return codes.

To help facilitate this, we can add some functionality to the `MTAEDA.Core` project. Let's open the `MTAEDA.sln` file in the `src` directory and add a new folder under Domain titled `Operations`. Within that folder, we will create two classes: `OperationResult` and `HttpOperationResult`. `OperationResult` will be an abstract class and be implemented by `HttpOperationResult`. The intent here is to encapsulate common return messaging along with the actual payload of the operation itself, where applicable. The `HttpOperationResult` class will allow us to further define return messaging based on standard HTTP return codes:

1. Let's start by adding the following code to the `OperationResult` class file:

   ```
   public abstract class OperationResult<T>
   {
       public abstract T? Value { get; set; }
       public Exception? Exception { get; set; }
       public bool IsSuccessful { get =>
           this.Exception == null; }
       public OperationResult() {


       }
   }
   ```

2. Declaring the `Value` property as `abstract` forces any class inheriting from the base to implement that property. This can be useful in instances where you may require additional control or processing around that payload value.

3. Next, add the following code to the `HttpOperationResult` class file to get started:

```
public class HttpOperationResult<T> :
  OperationResult<T>
{
    public override T? Value { get; set; }
    public string StatusMessage { get;
      private set; }
    public HttpOperationResult() {
        StatusMessage = Resources.SuccessMessage;
    }


    public static HttpOperationResult<T> Create()
    {
        return new HttpOperationResult<T> {
          StatusMessage = Resources.SuccessMessage
            };
    }


    public static HttpOperationResult<T> Create(T?
      value)
    {
        var operationResult = Create();
        operationResult.Value = value;
        return operationResult;
    }
```

4. Notice we've added a static `Create` method to this class with an overload for enabling the value of the operation payload to be passed in. The `StatusMessage` property will pull from a centralized resources file to keep verbiage concise and allow for a simpler way to ensure consistency. This doesn't set us up for returning multiple return message values, however. Let's add two more overloads that allow us to pass in an `HttpStatusCode` (from the `System.Net` library) and an `Exception` object:

```
public static HttpOperationResult<T> Create(T?
  value, HttpStatusCode statusCode) {
if((int)statusCode >= 400 && (int)statusCode
  < 600)
    {
```

```
            throw new InvalidOperationException
                ("If the HTTP status code is between
                400 and 600, a corresponding
                    Exception object should also be
                        set.");
        }
        var operationResult = Create(value);
        operationResult.StatusCode = statusCode;
        return operationResult;
    }


    public static HttpOperationResult<T> Create(T?
        value, HttpStatusCode statusCode, Exception
        exception)
    {
        var operationResult = Create(value,
            statusCode);
        operationResult.Exception = exception;
        return operationResult;
    }
```

5. Our final addition to this class will be a method to pull a static string from the Resources.resx file for the library and return it, depending on the HttpStatusCode presented. Add one more method to the class, as follows:

```
    private static string SetStatusMessage
        (HttpStatusCode statusCode)
    {
        int statusCodeValue = (int)statusCode;
        string statusMessage = Resources
            .SuccessMessage;
        switch(statusCodeValue)
        {
            case 401:
                statusMessage = Resources
                .BadAuthenticationMessage; break;
            case 403:
                statusMessage = Resources
```

```
                    .BadPermissionsMessage; break;
          case 500:
          case 501:
          case 502:
          case 503:
          case 504:
              statusMessage = Resources
                  .GeneralErrorMessage; break;
          default:
              statusMessage = Resources
                  .SuccessMessage; break;
      }
      return statusMessage;
  }
```

6. This `private` method can be adjusted to include other status codes depending on your needs or depending on what you'd like to test out. For the purposes of our example, we've left it brief to help illustrate the overall pattern of setting the status message. To use the new class, you can change the return value of the `RaiseEvent` method in `BaseCommandController` for the equipment domain from `IActionResult` to `HttpOperationResult`. When doing so, be sure to also change the method body to ensure the right object is being returned.

In order to drive notifications to the appropriate user interface, services such as **SignalR** can be used to broker those communications. There are other implementations within various **JavaScript** libraries that also facilitate notifications, including callbacks and **WebSockets**. While we are focused on the inner workings of the events and services, you may wish to experiment with different UI-based frameworks to put together an example that would display notifications back to the user.

We've seen how purpose-driven notifications are important to system continuity as well as to user experience. Having finished the last part of this chapter, we can now look forward to capturing information while each piece of the platform is doing its job.

Summary

In this chapter, we've explored the world of application resiliency, cloud-native design patterns, information redundancy to support business continuity, and the importance of clear and graceful communication. Using the Polly library, we have seen how different policies can be applied to method execution to promote resilience.

By using purpose-built services, we've seen how enabling event redundancy at each station can help protect information from becoming lost in the event of a communications failure. Also, we have seen how graceful service degradation can be augmented by graceful communication patterns, enabling user-centric feedback when things go amiss. As we look forward to *Chapter 13, Telemetry Capture and Integration*, we will take what we have learned with respect to resiliency and redundancy and augment it with telemetry capture to properly catalog information that can be used by operational teams later in the event of a platform issue.

Questions

1. What is the difference between resiliency, redundancy, and reliability?
2. What are the differences between the fallback and circuit breaker patterns?
3. What is the primary benefit of implementing a retry policy?
4. What is the difference between business continuity and disaster recovery?
5. For the example of the republishing service at the station level, what other options for executing that logic might exist to ensure events are published to the right topics?

Further reading

- *Resiliency and High Availability in Microservices* by Microsoft, available at https://learn.microsoft.com/en-us/dotnet/architecture/ microservices/architect-microservice-container- applications/resilient-high-availability-microservices
- *Polly.Net* by Polly Project, available at https://github.com/App-vNext/ Polly
- *Simmy, a chaos testing library for Polly.Net* by Polly Project, available at https:// github.com/Polly-Contrib/Simmy
- *Business Continuity and Disaster Recovery* by Microsoft, available at https:// learn.microsoft.com/en-us/azure/cloud-adoption-framework/ ready/landing-zone/design-area/management-business- continuity-disaster-recovery
- *Apache Kafka Connect* by Apache Software Foundation, available at https:// kafka.apache.org/documentation/#connect
- *Integrate Apache Kafka Connect support on Azure Event Hubs* by Microsoft, available at https://learn.microsoft.com/en-us/azure/event-hubs/ event-hubs-kafka-connect-tutorial

13
Telemetry Capture and Integration

In modern applications, collecting and acting on the information made available by all layers of the application is a requirement for determining areas for improvement, root-cause analysis when things go wrong, and even capturing customized information to help drive insights into business-specific workflows and activities.

In this chapter, we will examine the different options for capturing application-level and service-level telemetry, and how to ensure that relevant information is captured without producing unnecessary noise or overhead. You will also learn how to pinpoint meaningful telemetry and aggregate that as opposed to aggregating everything, which can lead to confusion, large storage footprints, and distrust of the information captured.

By the end of this chapter, you will be able to do the following:

- Identify the different types and levels of telemetry at the application, service, and component layers.

- Learn to implement basic telemetry capture using **OpenTelemetry** meter providers.

- Learn to implement custom telemetry capture using **OpenTelemetry** and .NET diagnostic meters.

- Determine what information is important enough to bubble up to monitoring agents and what level of alerting is required for each.

Technical requirements

You will find all the code examples for this chapter available in the folder for this chapter on GitHub here: `https://github.com/PacktPublishing/Implementing-Event-Driven-Microservices-Architecture-in-.NET-7/tree/main/chapter13` and `https://github.com/PacktPublishing/Implementing-Event-Driven-Microservices-Architecture-in-.NET-7/tree/main/src`.

> **Important note**
> The links to all the white papers and other sources mentioned in the chapter are provided in the *Further reading* section towards the end of the chapter.

Application- versus service- versus component-level telemetry

As covered in *Chapter 7, Microservice Observability*, we can collect simple telemetry by installing the **Application Insights** library as a dependency and configuring it to point to an existing Application Insights instance in Azure. This covers us at the code level for any immediate metrics, logging, or tracing that Application Insights can provide out of the box. While this is useful, we may need to break that down further and capture some specific baselines at various levels.

For our sample application, we would expect to capture telemetry at three different levels:

1. **Application** – Telemetry at the application level would include the overall health of the application, as well as any critical failures and the ability to troubleshoot distributed services. This may also include infrastructure- or cloud-platform-related status indicators.

2. **Service** – Telemetry at the service level would include measurements related to performance, faults, exceptions, or domain-specific telemetry where appropriate.

3. **Component** – Telemetry at the component level would be captured by the integration with Application Insights at the source code level or through custom code using the Application Insights SDK.

Another area in which things can become complicated is that of **Azure Monitor** and specifically **Container Insights**. Container Insights does offer a great deal of information at the cluster, namespace, deployment, service, and pod levels. It also collects a lot of information and can cause ingestion and data storage costs to increase considerably if proper retention periods are not set. When creating a new **Kubernetes** cluster in Azure, you can now specify whether you wish to install Container Insights or leave it turned off.

This is where frameworks such as OpenTelemetry can come in handy since it's purposefully built to be service- and platform-agnostic but offers a common way to expose the application telemetry to external systems. Let's dig into OpenTelemetry a bit further through an example illustrating how to easily set up and use telemetry capture within a standard API project.

Implementing non-intrusive telemetry capture

The idea of implementing baseline telemetry without the incurred overhead of adding custom code was not something that many deemed possible years ago. With the advent of platforms such as **New Relic**, **DynaTrace**, **DataDog**, and **Application Insights**, simply adding a library reference to your project and configuring connection information can enable a good amount of service- and component-level telemetry without further configuration. Even with a small amount of configuration and coding, these platforms can capture and expose metrics, logs, and traces.

Options exist beyond those platforms to capture telemetry as well. A question that you may find yourself asking is, "why would we want to implement this instead of using Application Insights?" On the one hand, you may want to gather this information and feed it to a platform tool within your Kubernetes cluster. You may also want more detailed control over what exactly is captured to avoid issues with bloated storage accounts or other storage media. There may be certain custom metrics that are easier to implement directly in .NET as opposed to implementing them using an **Application Performance Management** (APM) system's SDK and then working additionally to format that information within the APM platform itself. Having options for implementing telemetry and reporting on it can be beneficial depending on your use case.

Features within .NET and .NET Core allow us to leverage built-in means of measuring the items we really want. To get started, we'll look at three libraries that can add some additional value to component and service telemetry. One is the `System.Diagnostics.DiagnosticSource` library in .NET 7, and the other two are the OpenTelemetry meter SDKs for **Prometheus** and **Jaeger**. The OpenTelemetry framework, endorsed by the **Cloud Native Computing Foundation** (CNCF), is an open standard for producing and capturing values that can then be exported to reporting systems for further analysis or alerting. The SDKs for Jaeger and Prometheus demonstrate an extensibility that allows developers to integrate them with common Kubernetes platform components.

Within the `System.Diagnostics.DiagnosticSource` library, several object types can capture information. For our example, we'll focus on two significant object types:

- `Meter` – This class allows you to build out a collection of different counters that can be exposed using one of the meter SDKs listed previously.

- `Counter<T>` – This class allows you to create a counter object, which captures information of the `T` type for aggregation and reporting. The type constraint for `T` is that it must be a non-nullable value type.

`Meter` objects have a factory method, which can create a variety of different counters. The standard syntax for creating a `Meter` object and an associated counter would look like the following:

```
using OpenTelemetry.Metrics;
using OpenTelemetry;
using System.Diagnostics.Metrics;

var meter = new Meter("NameOfMeter");
var counter = meter.CreateCounter<int>(Name: "requests-
  received", Unit: "requests", Description: "Number of
    requests the service receives");
```

Next, we will turn to the implementation of more customized telemetry capture by updating our source code.

Implementing custom telemetry capture

We will create a simple example that will allow us to not only see how the OpenTelemetry SDK works but also how `Meter` objects can be used to house many different counters and data points:

1. Create a new Minimal API project in Visual Studio or with the `dotnet` command line. Here is the screenshot for the new Minimal API project:

Figure 13.1 – New project dialog in Visual Studio

Here is the screenshot for the `dotnet` command-line interface:

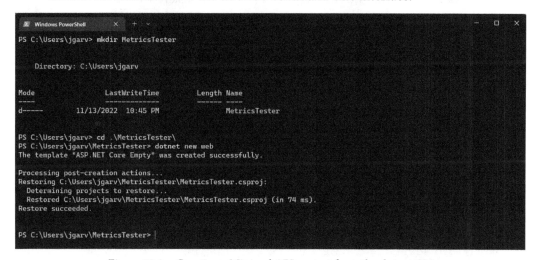

Figure 13.2 – Creating a Minimal API project from the dotnet CLI

2. Be sure to uncheck the **Use controllers** checkbox to use minimal APIs if using the new project dialog from Visual Studio:

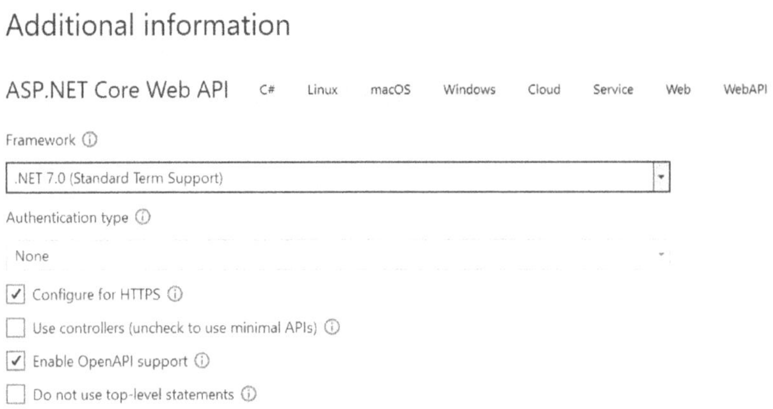

Additional information

ASP.NET Core Web API   C#   Linux   macOS   Windows   Cloud   Service   Web   WebAPI

Framework ⓘ

| .NET 7.0 (Standard Term Support) | ▾ |

Authentication type ⓘ

| None | ▾ |

☑ Configure for HTTPS ⓘ

☐ Use controllers (uncheck to use minimal APIs) ⓘ

☑ Enable OpenAPI support ⓘ

☐ Do not use top-level statements ⓘ

Figure 13.3 – Additional Information page in the New Project dialog

3. Add a NuGet reference to `System.Diagnostics.DiagnosticSource` and `OpenTelemetry.Instrumentation.EventCounters`:

```
dotnet add package System.Diagnostics.DiagnosticSource
dotnet add package OpenTelemetry.Instrumentation
  .EventCounters --prerelease
```

4. You may choose to leave the boilerplate service code in place (the `WeatherForecast` service) or replace it with your own. For this example, we will be using the boilerplate code.

5. Place the following declarations toward the top of the `Program.cs` file to create a new `Meter` object, as well as two counters to be used for collecting specific information:

```
var meter = new Meter("SampleMeter");
var counter = meter.CreateCounter<int>("requests-
  received", "Requests", "Simple counter using the
    Sample Meter");
var forecasts = meter.CreateCounter<int>("forecasts",
  "degrees", "Sample forecast captured from service
    call");
```

6. Create two new variables, one for instantiating the `SingleRandomObject` class and one for creating a new `MeterProvider` instance to expose it as a metric source (see the example code for this chapter for the definition of the `SingleRandomObject` class):

```
var singularity = new SingleRandomThing(314159);
MeterProvider? meterProvider = Sdk.CreateMeter
   ProviderBuilder().AddMeter("SampleMeter")
.Build();
```

7. Add the corresponding variables to the `Services` section of the application builder to allow for the `MeterProvider` instance to be used as a singleton, and the `SingleRandomObject` instance to be used as a transient:

```
builder.Services.AddSingleton(meterProvider);
builder.Services.AddTransient((c) =>
{
    return singularity;
});
```

8. Update the application so that a value is added to the requests received counter every time an action occurs. This will aggregate over time as the application runs. Keep in mind that restarting or stopping the application will reset the meter and counters:

```
app.Use((context, next) =>
{
    counter.Add(1, KeyValuePair.Create<string,
       object?>("path", (
         SourceUrl: context.Request.Path.Value,
         context.Request.Headers,
         ContentType: context.Request.ContentType?
            .ToString()
    )));
    return next(context);
});
```

9. Add the following line to the boilerplate code for the `GetForecast` method to capture information about the returned `forecast` objects. The key can be changed to any unique value you choose, such as a GUID. You can also convert the object being passed into JSON so that the full value is captured. This example only uses these arguments to the `Add` method to show an example of how to add information to the counter:

```
forecasts.Add(1, new KeyValuePair<string,
```

```
object?>(forecast[0].Date.ToShortDateString(),
    forecast[0]));
```

10. Add a new mapping for a GET method that can return different types of information based on a random number generator. This will give you the ability to see how dependency information is captured during the course of a RESTful call:

```
app.MapGet("/metricstest", () =>
{
    var nbr = singularity.GetRandomThing();
    if(nbr % 12 == 0)
    {
        var request = new HttpClient().GetStringAsync
            ("https://www.google.com/search?q=
                opentelemetry");
        return request.Result;
    }
    return $"Random number is {nbr}.";

}).WithOpenApi().WithName("MetricsTest");
```

11. Install the dotnet-counters command-line utility, which will present a console-based view of any meters and counters within a running process:

    dotnet tool install --global dotnet-counters

12. Start debugging the application by trying out the methods listed on the Swagger page that loads when the project starts.

13. Using the dotnet-counters utility, find the appropriate process ID for the sample project and run the monitor command with the process ID as the argument. Also, inspect the new Meter object and its output as described by the counter you created.

This example walked through how to easily wire up custom telemetry into a simple web API project. Next, let's investigate how to curate that custom telemetry and make it available to platform tools such as those commonly used by applications running in Kubernetes.

Bubbling up meaningful information

As we've seen, capturing both out-of-the-box and custom telemetry can be done with relatively little effort. The real effort is paring down that information to ensure there is not too much noise, and that relevant information on the health and performance of an application is what gets brought to the top. This will save not only the patience of those who monitor the application but can also result in a lowered cost in storage and processing by limiting the volume of information stored.

To help illustrate, we will be using two examples of exporters that are compatible with OpenTelemetry – Prometheus and Jaeger:

- Prometheus is a popular time-series database commonly used with Kubernetes to capture both standard and custom telemetry for applications. Information is typically scraped (read from an endpoint) into the Prometheus database and retained for a finite period. This information can then be used in trend analysis to identify patterns around the health, performance, or stability of an application or platform. Visualization tools such as **Grafana** can use Prometheus as a data source to facilitate the creation of dashboards and monitors, depending on what needs to be seen for that application.

- Jaeger is a distributed tracing framework that allows you to measure events across components (spans) and compose a view of those events in an end-to-end transaction flow (trace). This can be helpful, especially when troubleshooting a distributed application composed of many microservices. Span and trace data are usually stored in either **Elasticsearch** or **Cassandra** databases, although other sources can be used if you are so inclined.

Before starting the walkthroughs for the Prometheus and Jaeger exporters, please be sure to either have a Kubernetes cluster set up in the cloud or locally using **MiniKube**, **Docker Desktop**, or another local distribution of your choice. Please also ensure you have set the appropriate context so any commands issued using kubectl will affect the right instance. Also please be sure to have **Helm** installed, as it simplifies the installation of both components. The installation instructions are linked in the *Further reading* section.

Metric ingestion with the Prometheus exporter

Enabling and leveraging the metric ingestion using the Prometheus exporter for OpenTelemetry is not terribly complicated. To get things set up to use Helm, add the community repository for Prometheus by running the following commands:

```
helm repo add prometheus-community https://prometheus-
   community.github.io/helm-charts
helm repo update
```

Once the repository list has been refreshed, continue with the following steps to configure Prometheus metrics for your code:

1. Install Prometheus into your cluster using Helm and `kubectl` by entering the following line:

    ```
    helm install stable prometheus-community/kube-
        prometheus-stack -n monitoring
    ```

    Note there is a namespace specified at the end of the command—you may specify a namespace if you wish or omit it. Namespaces do help to keep similar services grouped together, however.

2. Add a reference to the `OpenTelemetry.Exporters.Prometheus` library using NuGet.

3. Add the code for the meter provider with the Prometheus exporter added:

    ```
    MeterProvider? meterProvider = Sdk.CreateMeter
        ProviderBuilder().AddMeter("SampleMeter")
            .AddPrometheusExporter().Build();
    ```

4. Add the code to expose the scraping endpoint to Prometheus:

    ```
    app.UseOpenTelemetryPrometheusScrapingEndpoint();
    ```

5. Run the program in Debug and test out making calls to the endpoints using the Swagger API page. Try each of the methods a few times to generate some information and then in a separate tab, navigate to `https://localhost:7182/metrics` to view the metric data being published by the service. *Figure 13.4* illustrates the output you should see when viewing the scraping endpoint:

```
# TYPE requests_received_Requests counter
# UNIT requests_received_Requests Requests
# HELP requests_received_Requests Simple counter using the Sample Meter
requests_received_Requests{path="(/metrics,
Microsoft.AspNetCore.Server.Kestrel.Core.Internal.Http.HttpRequestHeaders, )"} 1 1674415134940
requests_received_Requests{path="(/favicon.ico,
Microsoft.AspNetCore.Server.Kestrel.Core.Internal.Http.HttpRequestHeaders, )"} 1 1674415134940
requests_received_Requests{path="(/weatherforecast,
Microsoft.AspNetCore.Server.Kestrel.Core.Internal.Http.HttpRequestHeaders, )"} 10 1674415134940
requests_received_Requests{path="(/metrics,
Microsoft.AspNetCore.Server.Kestrel.Core.Internal.Http.HttpRequestHeaders, )"} 1 1674415134940
requests_received_Requests{path="(/metricstest,
Microsoft.AspNetCore.Server.Kestrel.Core.Internal.Http.HttpRequestHeaders, )"} 6 1674415134940
requests_received_Requests{path="(/metrics,
Microsoft.AspNetCore.Server.Kestrel.Core.Internal.Http.HttpRequestHeaders, )"} 1 1674415134940

# TYPE forecasts_degrees counter
# UNIT forecasts_degrees degrees
# HELP forecasts_degrees Sample forecast captured from service call
forecasts_degrees{_1_23_2023="WeatherForecast { Date = 1/23/2023, TemperatureC = 9, Summary = Sweltering,
TemperatureF = 48 }"} 1 1674415134940
forecasts_degrees{_1_23_2023="WeatherForecast { Date = 1/23/2023, TemperatureC = 5, Summary = Bracing, TemperatureF
= 40 }"} 1 1674415134940
forecasts_degrees{_1_23_2023="WeatherForecast { Date = 1/23/2023, TemperatureC = 48, Summary = Sweltering,
TemperatureF = 118 }"} 1 1674415134940
forecasts_degrees{_1_23_2023="WeatherForecast { Date = 1/23/2023, TemperatureC = 46, Summary = Bracing, TemperatureF
= 114 }"} 1 1674415134940
forecasts_degrees{_1_23_2023="WeatherForecast { Date = 1/23/2023, TemperatureC = 27, Summary = Cool, TemperatureF =
80 }"} 1 1674415134940
forecasts_degrees{_1_23_2023="WeatherForecast { Date = 1/23/2023, TemperatureC = 33, Summary = Balmy, TemperatureF =
91 }"} 1 1674415134940
forecasts_degrees{_1_23_2023="WeatherForecast { Date = 1/23/2023, TemperatureC = 41, Summary = Scorching,
TemperatureF = 105 }"} 1 1674415134940
forecasts_degrees{_1_23_2023="WeatherForecast { Date = 1/23/2023, TemperatureC = 2, Summary = Sweltering,
TemperatureF = 35 }"} 1 1674415134940
forecasts_degrees{_1_23_2023="WeatherForecast { Date = 1/23/2023, TemperatureC = 54, Summary = Warm, TemperatureF =
129 }"} 1 1674415134940
forecasts_degrees{_1_23_2023="WeatherForecast { Date = 1/23/2023, TemperatureC = 29, Summary = Scorching,
TemperatureF = 84 }"} 1 1674415134940

# EOF
```

Figure 13.4 – Sample display of Prometheus metrics published

6. Configure Prometheus to scrape the newly exposed metrics endpoint by creating an entry in the Prometheus configuration.

After debugging and testing, you may wish to adjust the counter types or add different types of objects to the meter, such as histograms. After you've finished experimenting, it is time to move on to examine the setup and usage of the Jaeger exporter.

Distributed tracing using the Jaeger exporter

Enabling the Jaeger exporter with OpenTelemetry is also a straightforward endeavor. To get things set up to use Helm, add the community repo for Jaeger by running the following commands:

```
helm repo add jaegertracing https://jaegertracing.github.io
/helm-charts
helm repo update
```

Once the repository list has been refreshed, continue with the following steps to configure distributed tracing with Jaeger for your code:

1. Install Jaeger into your cluster using Helm by entering the following at the command line:

   ```
   helm install jaeger jaegertracing/jaeger -n monitoring
   ```

2. Add a reference to the OpenTelemetry.Exporters.Jaeger library to your project via NuGet.

3. Also, add the following OpenTelemetry packages to enable console monitoring, as well as ASP.NET Core instrumentation:

   ```
   dotnet add package OpenTelemetry --prerelease
   dotnet add package OpenTelemetry.Exporter.Console --
     prerelease
   dotnet add package OpenTelemetry.Extensions.Hosting --
     prerelease
   dotnet add package OpenTelemetry.Instrumentation
     .AspNetCore --prerelease
   ```

4. Add the code for the tracer provider below the code for MeterProvider and use the OpenTelemetry.Sdk object to create a provider using the Jaeger exporter:

   ```
   TracerProvider? traceProvider = Sdk.CreateTracer
     ProviderBuilder().AddSource("SampleMeter")
       .AddJaegerExporter().Build();
   ```

5. Configure the builder context to use tracing as well by updating the Services collection with a new configuration:

   ```
   builder.Services.AddOpenTelemetry().WithTracing
     (config => {
       config.AddJaegerExporter();
       config.AddAspNetCoreInstrumentation();
       config.AddConsoleExporter();}).StartWithHost();
   ```

251

BubBubbling up meaningful information251Bubbling up meaningful information   251

6. Add a singleton instance to the `Services` collection, which will reference a
   `Tracer` object to be used with a new utility class, `SimpleActivityClass`:

   ```
   builder.Services.AddSingleton<Tracer>(traceProvider
       .GetTracer("SampleTracer"));
   ```

7. Review the class code for `SimpleActivityClass` to see how we will be
   implementing both the `Activity` model and the `TelemetrySpan` object to
   capture the tracing information. Once you are finished, add the following new
   method to the `Program.cs` file:

   ```
   app.MapGet("/activity", () =>
   {
       using (var act = new SimpleActivityClass(app
           .Services.GetRequiredService<Tracer>()))
       {
           act.DoSomethingTraceable();
       }
       return "ok";
   }).WithName("TraceActivity");
   ```

8. Run the project in debug mode and send some sample requests using each of the
   methods. You should see trace information showing up in the console while the
   program is running.

9. Package and deploy the program to your Kubernetes cluster in a namespace of your
   choosing.

10. Run a `port-forward` command using `kubectl` to expose the service instance
    locally. Once completed, you can use the URL provided to send requests to the API
    directly, using the base URL plus the relative path to the method. For example, if the
    `port-forward` command maps the service to `https://localhost:37110/`,
    you can add `api/activity` to the end of that URL to call the `Activity` method
    directly.

11. Submit a few different method calls to generate some traffic. Once you are
    satisfied, run a `port-forward` command using `kubectl` to bring up the Jaeger
    dashboard. *Figure 13.5* illustrates a sample of the Jaeger dashboard's landing page:

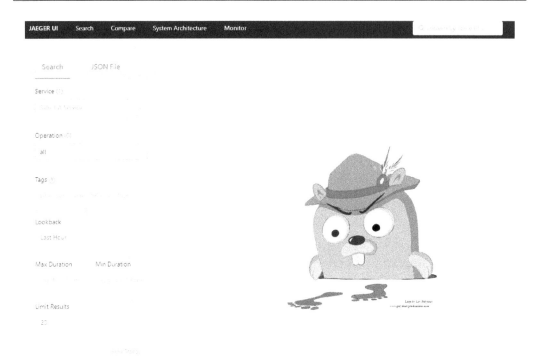

Figure 13.5 – Jaeger dashboard landing page

12. You should now see the Jaeger dashboard and be able to search for traces related to your newly deployed service, along with any traces captured.

With configurations now in place for two common platform tools used by engineers who work with Kubernetes-based applications, you've ensured that capturing consistent telemetry can be done in any of the services in the sample application.

Summary

In this chapter, we explored different types of telemetry, including the common areas where that information can be generated, captured, and leveraged. We saw how implementing custom meters and counters using .NET 7 can be relatively simple and provides value. Through the use of the OpenTelemetry framework, we learned how to not only create and expose custom telemetry but also saw how it can be integrated into standardized platform tools such as Jaeger and Prometheus to assist with monitoring, alerting, and tracing across our event-driven system.

In our last chapter, *Chapter 14, Observability Revisited*, we will take a deeper look into how APIs and microservices can be observed further using concepts such as service discovery and metadata cataloging.

Questions

1. In what cases would you not rely solely on platform telemetry services such as Application Insights to capture and analyze information from your services?

2. What's the primary difference between the `Activity` and `Meter` object types in .NET?

3. Are there any considerations to keep in mind when capturing telemetry, such as platform constraints, cost, or performance?

Further reading

- *Collect metrics* by Microsoft, available at `https://learn.microsoft.com/en-us/dotnet/core/diagnostics/metrics-collection`

- *Distributed tracing instrumentation walkthroughs* by Microsoft, available at `https://learn.microsoft.com/en-us/dotnet/core/diagnostics/distributed-tracing-instrumentation-walkthroughs`

- *Metric APIs comparison* by Microsoft, available at `https://learn.microsoft.com/en-us/dotnet/core/diagnostics/compare-metric-apis`

- *Lens/The Kubernetes IDE* by Mirantis Inc., available at `https://k8slens.dev/`

- *Installing Helm* by the Helm authors, available at `https://helm.sh/docs/intro/install/`

- *What is Prometheus?* by the Prometheus authors, available at `https://prometheus.io/docs/introduction/overview/`

- *Jaeger: Getting Started* by the Jaeger authors, available at `https://www.jaegertracing.io/docs/1.41/getting-started/`

- *Collect a Distributed Trace* by Microsoft, available at `https://learn.microsoft.com/en-us/dotnet/core/diagnostics/distributed-tracing-collection-walkthroughs`

- *.NET distributed tracing* by Microsoft, available at `https://learn.microsoft.com/en-us/dotnet/core/diagnostics/distributed-tracing?source=recommendations`

- An OpenTelemetry visualization dashboard by *Monitoring Artist*, available via GrafanaLabs at `https://grafana.com/grafana/dashboards/15983`

14
Observability Revisited

In *Chapter 7*, *Microservice Observability*, we looked at how to enable the internal observability of microservices. We examined how to aggregate metrics, logs, and traces so that we can efficiently observe and understand performance and problems across a distributed, highly scalable solution.

In this chapter, we will look at how to make our services externally observable to others. You will learn about methodologies for publishing service metadata to your organization, cataloging and versioning microservice metadata, and how to promote the discovery of shared services within an organization.

By the end of this chapter, you will be able to do the following:

- Understand how to expose service information with service metadata
- Understand how to leverage service discovery and implement it using Consul
- Learn about service publishing and understand the supporting features of Azure API Management
- Learn about the Ortelius service offering and how it can catalog, track, and trace dependencies for services

Technical requirements

You can find all the code examples for this chapter in the relevant folder on GitHub, which is available at `https://github.com/PacktPublishing/Implementing-Event-Driven-Microservices-Architecture-in-.NET-7/tree/main/chapter14`.

> **Important note**
>
> The links to all the white papers and other sources mentioned in this chapter are provided in the *Further reading* section toward the end of the chapter.

Sharing API services

So far, in our example TAEDA application, we have focused on various domain information that is closely interrelated. As a lone developer, you can probably recall most of the API endpoints you created in one microservice when you come to utilize them in another microservice. However, in practice, we don't develop in isolation. One of the key benefits of microservice architecture is the ability to divide and conquer by assigning smaller domains to different teams (or lone developers). In fact, we don't want to limit the understanding of our APIs to just the teams associated with the main solution, but to any other teams that might be able to benefit from these same services.

Let's imagine a scenario where there is an intent to conduct a study on transit foot traffic patterns and how they correlate with major sporting events around various cities. We already have many services designed to support the backbone of the transit management system. There is bound to be a level of reuse for the purposes of this study. Accessing those services and capabilities should not depend on institutional knowledge and scanning through piles of documentation and code. These services should be easily discoverable, well-defined, and published.

In the same way that Excel spreadsheets and local Access databases gave way to data lakes and warehouses, server-based monolithic applications give way to sharable distributed services and business capabilities.

The question is… how can we get our APIs out there for others to understand and consume?

Let's start by making sure our services are well defined.

Generating service information

Generating service information is a critical prerequisite for exposing services to wider audiences. It not only serves the purpose of helping consumers understand and test service calls, but it also allows them to stay current with version changes in the service definitions.

It did not take long for the open source community and software industry, in general, to standardize what service information should look like. In parallel, the tooling that helps generate this service information has evolved. Before looking at *our* tooling of choice, first, let's understand the most commonly used API standard developed by the OpenAPI initiative.

OpenAPI specification

At the time of writing, the **OpenAPI Specification v3.1.0** is the currently published version. The version number itself isn't so important to our learning, but rather what the OpenAPI specification is and what it hopes to achieve.

The following is taken from the beginning of the OpenAPI specification, answering the question of *What is the OpenAPI Specification?*:

The OpenAPI Specification (OAS) defines a standard, programming language-agnostic interface description for HTTP APIs, which allows both humans and computers to discover and understand the capabilities of a service without requiring access to source code, additional documentation, or inspection of network traffic. When properly defined via OpenAPI, a consumer can understand and interact with the remote service with a minimal amount of implementation logic. Similar to what interface descriptions have done for lower-level programming, the OpenAPI Specification removes guesswork in calling a service.

In short, this says we should conform to describing what a service is and how you can call it in a very standard way, regardless of who you are, whom you work for, what development language you use, or the audience your services may serve.

Thanks to the quick evolution of tooling to support this standard, we have two distinct approaches to conform with the OpenAPI standard:

1. Design an API by describing it as an OpenAPI standard, and then generate code stubs for new services, or compare the existing service code against it.

2. Code an API to meet your needs, and then generate an OpenAPI specification that describes it.

There is plenty of tooling across most major development languages that support both approaches. The OpenAPI initiative recommends using a design-first approach as best practice. This would indicate that the tooling for describing an OpenAPI-compatible specification and then generating or testing against code is the best choice. However, be aware that the tooling that supports reverse engineering the specification from existing code is still of great value. While we might design the API first, and then generate new code or test existing code against it, the code is always at risk of change. So, having tooling that can generate the specification at runtime means we can always see as-is definitions. In effect, we want to statically align a desired API with actual code, but we also want to dynamically present the as-is API to any consumer at any given time – regardless of its alignment with a desired design.

Also, being honest about my concentration levels and over-eagerness to code, I do often write code first when creating new API services. There are some exceptions – mostly when I am developing large-scale systems that will go into production use. If you build a poor design because you are too head-deep in code, you are building liabilities that will come back and hurt. But in my spare time and for side projects, I jump straight to code and let the generator tooling serve me an API specification.

A brief history of Swagger

When developing in .NET Core, the most widely used OpenAPI specification generation tool is called **Swashbuckle**. This uses an object model and middleware called **Swagger**.

The OpenAPI specification originates from tooling called **SWAGR** (pronounced *Swagger*), which was created by developers at Wordnik (an online dictionary search service). Its purpose was to produce easy documentation for developers who would use the API and, in doing so, accelerate the monetization of its access.

Fast forward many years, and that specification has been since acquired and then donated to the open source community (via the Linux Foundation) and has spun off into a multitude of tooling compatible with the standard. As such, you will often find the names Swagger and OpenAPI used interchangeably.

When seeking out any related tooling, it is important to see what versions of the OpenAPI specification are supported. In general, you will find more tooling with Swagger in the name than OpenAPI, but these mostly support the latest OpenAPI specification.

Generating the Swagger documentation

We will demonstrate how to autogenerate an OpenAPI specification document, at runtime, for our producer service. This requires two different elements to be configured in our services:

1. The autogeneration of the specification in JSON format.

2. The presentation service will provide a UI to navigate the specification.

The initial step for both configurations is to add a reference from the Swashbuckle library to the producer project. In the folder containing the `producer.csproj` file, run the following command:

```
dotnet add package Swashbuckle.AspNetCore --version 6.5.0
```

To have the API code signatures detected for automatic documentation generation, we cannot use the inline API definitions as they currently exist in `Program.cs`. Instead, we must create API controller files and decorate them with attributes that make them discoverable to Swashbuckle.

Converting API methods from inline into controllers

We create a new file called `ProducerController.cs` and, within that, a method called `Send()`:

```
[Route("[controller]")]
    [ApiController]
    [ProducesResponseType(typeof(ProblemDetails), 400)]
    public class ProducerController : ControllerBase
    {
        private readonly HttpContext httpcontext;
        public ProducerController(IHttpContextAccessor
          httpContextAccessor)
        {
            this.httpcontext =
                httpContextAccessor.HttpContext;
        }

HttpPost("Send")]
        [ProducesResponseType(typeof(string), 200)]
        public async Task<IActionResult> Send([FromBody]
          string message)
        {
}
}
```

To define the URL path used to access the method, we set the `Route` decorator attribute on the class and the `HttpPost` decorator attribute on the method:

```
[Route("[controller]")]

...

[HttpPost("Send")]

...
```

The `[controller]` string argument in the `Route` attribute defines that the URL path for this controller should be taken from the filename. This value is a special setting that refers to a built-in naming template. .NET will drop the word *controller* from the filename and use that as the path. In our example, `ProducerController.cs` will result in `/producer/` as a path.

The `HttpPost` attribute defines that the API call is a **POST** method, and it is named `Send`. The full path to this API endpoint, as defined in the new controller file, will be `/producer/send`.

We add more attribute decorations that will result in metadata for the API signatures:

```
[ProducesResponseType(typeof(ProblemDetails), 400)]
[ProducesResponseType(typeof(string), 200)]
```

Note that `Microsoft.AspNetCore.Mvc.ProducesResponseType` is applied at the class level and the method level. At the class level, it acts as a default for all methods. So, we can identify that any HTTP `400` messages will return a response in the `Microsoft.AspNetCore.Mvc.ProblemDetails` structure. This is a standard class definition given that, in most cases, the HTTP `400` response format is handled internally. At the method level, we identify that an HTTP `200` (success) response will return a type of `string`.

Swashbuckle uses another .NET Core library called `ApiExplorer` to extract all this metadata that is being added to the code. To identify our class as one that should be indexed for metadata, we add the `ApiController` attribute.

The `Microsoft.AspNetCore.Http.HttpContext` is not immediately accessible as with the inline API definitions. To access this, the controller class uses dependency injection to obtain a reference to an instance of the `IHttpContextAccessor` interface. In the `ProducerController` constructor, we assign this to a private read-only field for use in the method body.

Now we can add the code within the `Send()` method. It follows the same logic as the inline call that we defined in `Program.cs` in *Chapter 2, The Producer-Consumer Pattern*, and within that, a method called `Send()`:

```
[HttpPost("Send")]
        [ProducesResponseType(typeof(string), 200)]
        public async Task<IActionResult> Send([FromBody]
          string message)
        {
            var causationId =
              System.Diagnostics.Activity.Current.TraceId
              .ToString();
            var correlationId =
              Request.Headers.RequestId.FirstOrDefault() ??
              causationId;
            var svc = httpcontext.RequestServices
              .GetRequiredService<IProducerService>();
            await svc.SetTopic("equipment");
            try
            {
                if (message == null)
                {
                    throw new InvalidDataException("Must
                      have a message body");
                }
                await svc.Send(message, correlationId,
                  causationId);
            }
            catch (Exception ex)
            {
                return new BadRequestObjectResult
                  (ex.Message);
            }
            return new OkObjectResult("Ok!");
        }
```

Finally, although we can access the **request body** from the HttpContext attribute, the controller definition allows us to be more explicit about the content expected in the call. This is done by simply adding the expected arguments to the method signature. As this call uses a POST method, we label the attribute as being FromBody. Note that we have kept this argument as a simple string, but it could easily be defined as a strongly typed class or struct that would be represented in JSON format in the POST body.

Adding startup services and middleware

The modifications we made in building out the controller have done all the work we need to generate a basic OpenAPI specification at runtime. However, there are several services to inject and middleware to configure for this to be completed.

In the `Program.cs` file, we add a few new services:

```
builder.Services.AddControllers();
builder.Services.AddMvcCore().AddApiExplorer();
builder.Services.AddSwaggerGen();
builder.Services.AddHttpContextAccessor();
```

The additions should be relatively self-explanatory. We need the `Controllers` service to support the use of individual controller file definitions. The `ApiExplorer` service produces metadata based on the controller class structure and various attribute decorations. The `SwaggerGen` service produces the Swagger JSON file in the latest OpenAPI format. And finally, the `HttpContextAccessor` service is produced so that we can access `HttpContext` via dependency injection.

Configuring the middleware follows a similar pattern that should be relatively self-explanatory:

```
app.UseSwagger();
app.UseSwaggerUI();
app.UseRouting();
app.UseEndpoints(endpoints => {
    endpoints.MapControllers();
});
```

Configuring the Swagger middleware will allow the Swagger JSON file to be accessible on a predefined path (`/swagger/v1/swagger.json`). Configuring the SwaggerUI middleware will enable a UI to be served on a predefined path (`/swagger/index.html`). Routing is required for the `Route` attribute in the controller definition to work. And finally, configuring the use of `endpoints` with `MapControllers` will ensure that the controllers are accessible on their configured path (`/producer/send`).

One thing that took me some time to understand, despite how obvious it feels now, is that `SwaggerUI` depends on the JSON file generated by the `Swagger` middleware. If you don't need an interactive UI, but you do need the OpenAPI specification to be served in JSON format, you only require the `Swagger` middleware and not `SwaggerUI`.

Viewing the runtime Swagger UI

Now when we build and run the producer code, everything should look and feel the same at startup. If you access the `/swagger/index.html` URL path, you will be presented with a clean UI that shows our API signatures and even lets you interact with them directly:

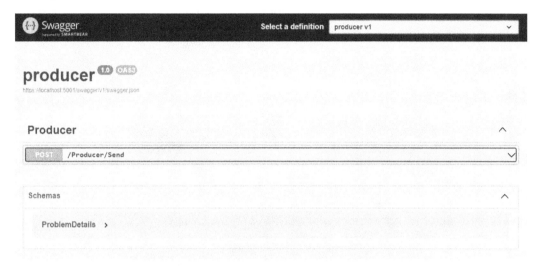

Figure 14.1 – Swagger UI

When you open the **/Producer/Send** UI accordion, there is an option to **Try It Out**, which will let you populate the **Request body** content and execute the API calls live:

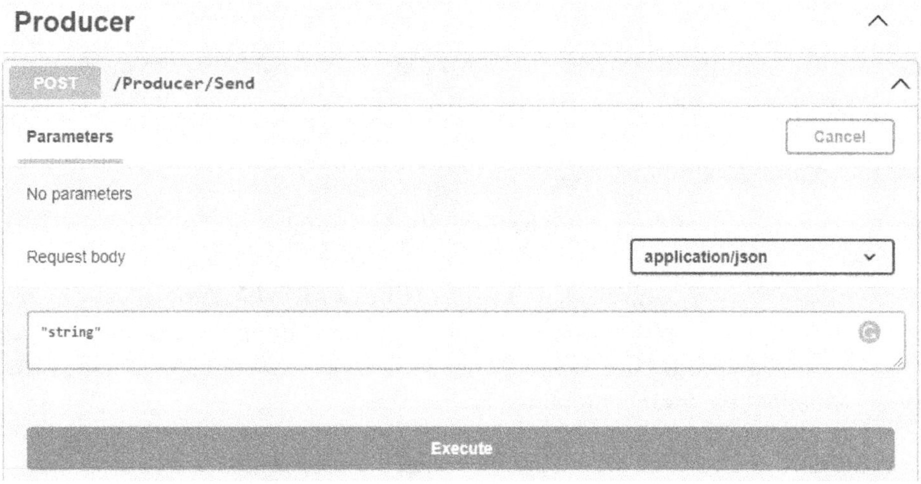

Producer

Figure 14.2 – "Try it out" for an API method

We even have schemas and examples for the potential response types, as we defined with the `ProducesResponseType` attributes:

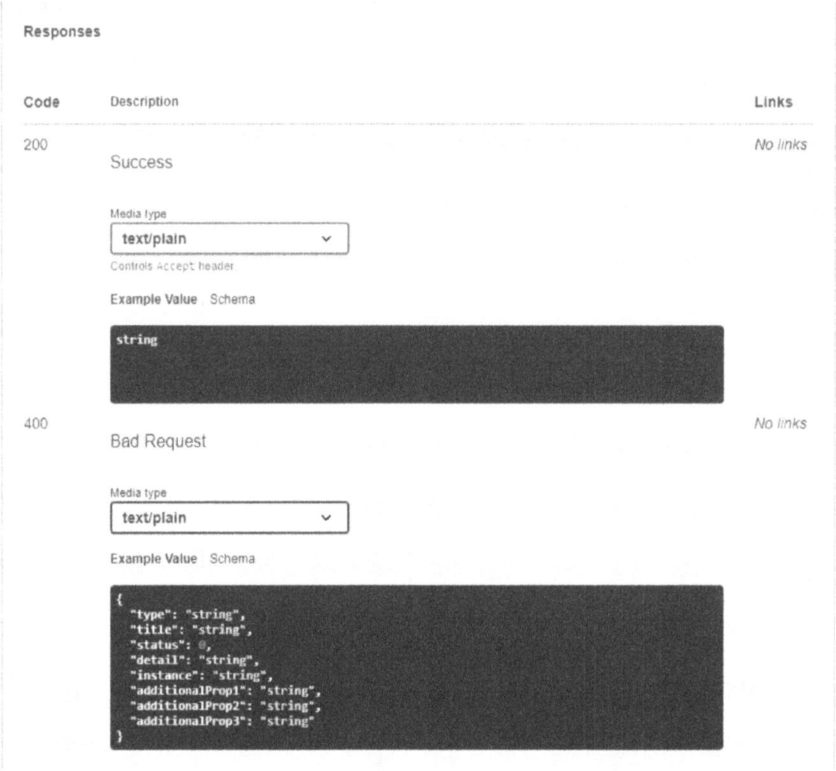

Figure 14.3 – The responses documentation for an API method

Now there are many things that can be customized and improved here with more code attribute decorations and signature patterns. For example, a description of the call, along with any other specific instructions, links to other documentation about the object schemas, authentication methods, and more.

You might have noticed that despite not having a version number in the URL path for our method, the UI shows our call as *producer v1*. The OpenAPI specification requires that all API definitions are version-controlled, so out-of-the-box v1 will be automatically applied. This can also be customized to the point where you can serve multiple versions of an API signature from the same code base. If you follow a semantic versioning pattern, this is a very useful element of the Swagger tooling.

Hopefully, this runtime UI capability immediately reinforces how auto-documenting interactive APIs can empower other developers to integrate with your services. However, this still isn't really the OpenAPI *specification*. This is a web page!

Viewing the runtime OpenAPI specification

At the top of the Swagger UI, there is a link to a JSON file. By default, it will end with /swagger/v1/swagger.json. Clicking on this will take you to the JSON content served by the Swagger (not SwaggerUI) middleware. This is the true OpenAPI specification format:

```
{
  "openapi": "3.0.1",
  "info": {
    "title": "producer",
    "version": "1.0"
  },
  "paths": {
    "/Producer/Send": {
      "post": {
        "tags": [
          "Producer"
        ],
        "requestBody": {
          "content": {
            "application/json": {
              "schema": {
                "type": "string"
              }
            }
```

Figure 14.4 – swagger/v1/swagger.json

> **Important note**
> Make sure that you save this file locally on your filesystem. We will use it again when we try out **Service Publishing** later in this chapter.

The format here is compliant with OpenAPI (**3.0.1**, in this example). Navigating the specification documentation, you can trace through which sections, subsections, and properties are required. You will find all of these in the document. This is why the Swagger middleware will automatically assign a version even if you don't configure one.

It is recommended to further enrich the generated API specification through further refinement of the code. While this can be time-consuming, it adds so much value to the many future consumers of your work. This is also a perfect moment to reflect on my personal preference to use code first instead of design first. There is also a desktop application and SaaS offering for SwaggerUI. In that edition, we visually design APIs first, as if building the interactive web page for runtime. When complete, it can generate code stubs in any supported language, which includes all the decorators and structures required for the various specification sections.

With a documented API following the OpenAPI specification, we need to make our API available for others to find. We will cover that in the *Service publishing* section. Before we reach publishing, there is another important part of service availability that we must cover: service discovery.

Service discovery

In the current context of *sharing APIs*, it's easy to think of service discovery as *a developer finding an API service to use*. For this chapter, that is more of a downstream event (pun intended) of service publishing. Instead, we are going to focus on service discovery in the context of seeking out available and healthy endpoints for our API.

When achieving the levels of scalability that event-driven architecture offers, we, of course, have many instances of our services running across the unpredictable landscape of the cloud. So, when we make one of our APIs available as a single unified service, it is almost a certainty that this is backed with multiple instances of the same runtime across almost any combination of infrastructure configurations. How do we make sure that requests to our API are directed to a healthy service instance?

It's not quite load balancing...

This might appear like a job for a load balancer. However, load balancers come with their own challenges.

Firstly, they are not designed for fast addition and removal of backend pool endpoints. This is a problem when we might have many different infrastructure components spinning up and shutting down instances of our services at a rapid pace.

Secondly, load balancers are not traditionally designed for high availability. They become a choke point for the service. If the load balancer fails, all services behind it fail.

Now cloud platform services are catching up on these problems – but only by adopting the same mechanics, as we will explore in service discovery. This means that the services are designed for rapid service registration and deregistration, and implemented with a peer-to-peer network of load-balancing services.

We will explore one of the leading solutions for service discovery: **Consul** by HashiCorp.

Exploring Consul

Consul is a service mesh and discovery platform that helps abstract the availability, reliability, and security aspects of how one service communicates with another. In a complete deployment, not only does it know where all instances of a service are and what their health status is, but it also handles the encrypted communication between them and can discover new services with no application code changes required.

For the scope of this chapter, we want to focus only on service discovery. Our desired end state is a highly available Consul cluster that is aware of all our API service instances and their health statuses and provides a way to address these services that abstracts any logic away from the calling application.

Setting up the Docker network

As we are using Docker containers with docker-compose, there are some limitations to how Consul can operate. The Consul cluster will eventually act as a **DNS server** so that services can be addressed using a predictable hostname. Consul will manage the DNS responses so that traffic is **balanced** across service instances and only **healthy instances** are addressed. In a full production configuration, DNS binding and forwarding would be used to integrate the Consul DNS into the network setup. For the purposes of this chapter, Consul will be the *only* DNS used by our service instances. To support this, we must assign static IP addresses to all the containers within the docker-compose configuration.

At the end of the docker-compose.yml file, we define a virtual network and address space:

```
networks:
  mtaeda:
    driver: bridge
    ipam:
      config:
        - subnet: 10.3.0.0/27
```

The network driver is called bridge and will ensure that there is a bridged route between your local host machine and the virtual network address space inside Docker.

Now we can assign each existing container in the configuration to this network with a static IP using the following template:

```
networks:
  mtaeda:
    ipv4_address: 10.3.0.x
```

Adding more service instances

For Consul to balance instances of the same service, we require more than one instance of each. We do this by simply duplicating the producer and consumer configurations in `docker-compose.yml`.

They require a different section name (*producer1*, *producer2*, and so on), unique IP addresses, and a different port mapping to the local machine.

Adding the Consul cluster

Normally, a Consul cluster takes the form of multiple server agents and client agents.

The servers provide the peer-to-peer network, with a leader election process, to manage the state of the registered services. This is similar, in nature, to how a Kafka cluster elects a leader, except that is controlled by an instance of Zookeeper. Consul uses a consensus protocol, called **Raft**, to allow the servers to elect a leader. There is a fantastic link to a visual explanation of the Raft protocol in the *Further reading* section.

The clients provide the proxy endpoints that are close to the service instances. These may be in a Kubernetes cluster, running on individual VMs, or in our case, just another Docker container.

We will implement a simple one-server, one-client configuration by adding the following to the `docker-compose.yml` file:

- The server configuration is as follows:

```
consul-server:
    image: hashicorp/consul:1.10.0
    container_name: consul-server
    hostname: consul-server
    restart: always
    volumes:
      - ./consul/server.json:/consul/config/server.json:ro
    ports:
      - "8500:8500"
      - "8600:53/tcp"
      - "8600:53/udp"
    networks:
      mtaeda:
        ipv4_address: 10.3.0.9
    command: "agent"
```

Here, we have defined `consul-server` as a container built on an image published by HashiCorp. We mount the local `server.json` file as a read-only file inside the container. This defines how the server should be configured when it starts, including the fact that it will operate as a server and not a client. We don't need to concern ourselves with the details of this configuration, as this file is taken directly from HashiCorp's example for Service Discovery. This is linked in the *Further reading* section if you wish to dig deeper.

- And the client configuration is as follows:

```
consul-client:
    image: hashicorp/consul:1.10.0
    container_name: consul-client
    hostname: consul-client
    restart: always
    volumes:
      - ./consul/client.json:/consul/config/client.json:ro
    ports:
      - "8601:53/tcp"
      - "8601:53/udp"
    networks:
      mtaeda:
        ipv4_address: 10.3.0.10
    command: "agent"
```

In the same way as the server, we have defined `consul-client` as a container built on the same image published by HashiCorp. We mount the local `client.json` file as a read-only file inside the container. This defines how the client should be configured when it starts.

Apart from the name and IP address, they both look very similar. The difference is in the configuration JSON files, which you will find in the `consul` folder.

Accessing the server UI

Now we can run `docker-compose up` and see both the Consul client and server running. Visiting `http://localhost:8500` in your browser will bring you to the Consul server UI. Under the services section, you will see **1 instance** of Consul (referring to the server) and **2 instances** of the Consul DNS (one on each server and client):

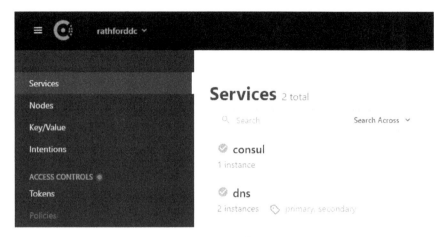

Figure 14.5 – Consul server UI

The service monitoring definitions for the DNS are in the JSON configuration files. The definition for server monitoring is created by default in any Consul cluster.

Now we have a basic Consul cluster running, we need to ensure it can track the multiple instances of our services.

Service discovery/registration

When using Consul in an environment such as Kubernetes, it's quite simple to add annotations to your pod specs to enable service registration. In this regard, Consul is *discovering* service instances as they are created, and removing them as they are destroyed. Pretty easy!

Running in the docker-compose environment, service registration is slightly more involved. Consul has an API endpoint for registering service instances. In Kubernetes, Consul deploys a controller that can call this endpoint as containers with the relevant pod specs are created. For docker-compose, we will need to call that API endpoint ourselves. To do this, follow these steps:

1. We can use a bash script to be called at our container's entrypoint:

```
#!/usr/bin/env bash
IP=`ip addr | grep -E 'eth0.*state UP' -A2 | tail -n 1 |
  awk '{print $2}' | cut -f1 -d '/'`
read -r -d '' MSG << EOM
{
  "id" : "consumer-$IP",
  "name": consumer",
```

```
        "address": "$IP",
        "port": 123,
        "check": {
            "tcp": "$IP:123",
            "interval": "5s"
        }
    }
    EOM
    curl -v -XPUT -d "$MSG" http://consul-client:8500/v1/
        agent/service/register && dotnet consumer.dll "$@"
```

The IP address of the container is extracted from the `ip addr` command. This is then used to build a PUT request body that will tell Consul that a new instance of a service has been created. The consumer service defines a health check in `TCP` `123`. This is the same liveness endpoint that we created in *Chapter 7*, *Microservice Observability*.

2. The producer script is almost identical, with changes in places you would expect them:

```
#!/usr/bin/env bash
IP=`ip addr | grep -E 'eth0.*state UP' -A2 | tail -n 1 |
    awk '{print $2}' | cut -f1 -d '/'`
read -r -d '' MSG << EOM
{
    "id" : "producer-$IP",
    "name": "producer",
    "address": "$IP",
    "port": 80,
    "check": {
        "http": "http://$IP:80/healthz",
        "interval": "5s"
    }
}
EOM

curl -v -XPUT -d "$MSG" http://consul-client:8500/v1/
    agent/service/register && dotnet producer.dll "$@"
```

Note that for the producer, the liveness endpoint created is `http` and the path is `/healthz`.

3. In the `docker-compose.yml` file, we can now override the container entrypoint to execute this script. At the same time, we will point each service instance to use the Consul server as its DNS using a static IP address:

    dns:

    **   - 10.3.0.9 #consul-server**

    entrypoint: ./entrypoint.sh

4. Now we can restart the environment with docker-compose down, followed by docker-compose up.

After a few seconds, the Consul server UI will show both the producer and consumer services, each having **2 instances**, and all being healthy:

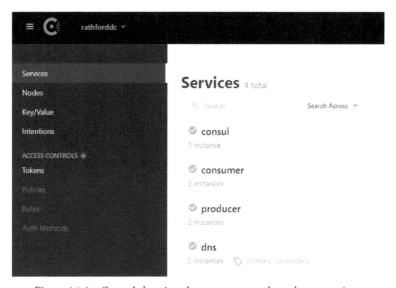

Figure 14.6 – Consul showing the consumer and producer services

At this point, we have instances of our services being actively monitored by Consul for health. Should we create any new instances of our services, they will self-register with Consul. Now we just need to make sure they resolve correctly.

Service resolution

Any developer who wishes to access our services just needs to use the Consul DNS resolver or a DNS server that binds or forwards to it. We have configured the producer and consumer instances to resolve DNS with the Consul server. By design, they should never really have to care about direct communication with each other. Their interaction is purely through events. However, for the purposes of demonstrating Consul service discovery, we can connect to a consumer instance and see how to reach out to a healthy, load-balanced instance of the producer:

1. Using the Docker desktop, connect to a Terminal session on one of the consumers. Use the dig command to query the Consul DNS for the producer.service.consul address:

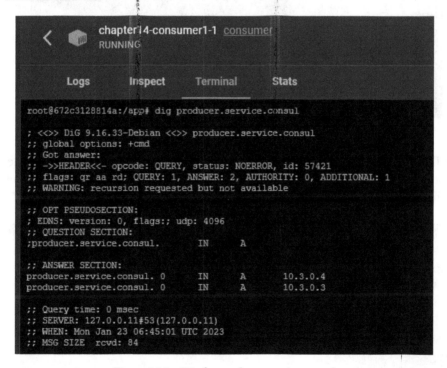

Figure 14.7 – Dig for producer.service.consul

You can see that the Consul DNS resolves this to the addresses of both producer instances. If you repeat this command a few more times, you will notice how the order of preference changes as Consul attempts to influence a level of load balancing.

2. Next, stop one instance of the producer and run the dig command again:

Figure 14.8 – Another Dig for producer.service.consul

Now the DNS entry only resolves to the single remaining healthy instance. Even if the other producer instance was still running, but the liveness endpoint was not reporting as healthy, the instance would be eliminated from the address resolution by Consul:

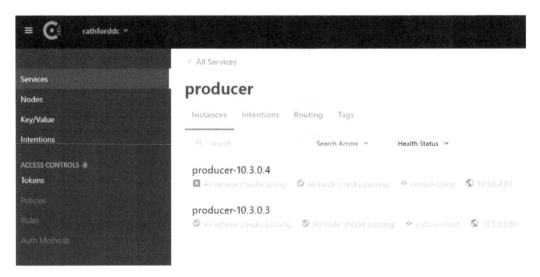

Figure 14.9 – Consul showing only one healthy instance of the producer service

Our use of service discovery is only part of what Consul has to offer. As a service mesh, Consul can facilitate communication between services by acting as a secure, policy-managed proxy transport. When configured, Consul DNS will instead return the appropriate address for its nearest client proxy and handle the routing and load balancing internally.

We have understood how to ensure our service instances are discoverable and available. And our autogenerated documentation is up to standard. Next, we need to publish our service so that others can find it.

Service publishing

Now that we have an API, an easy-to-navigate specification for it, and a service discovery layer that manages the distribution of requests to healthy service instances, developers just need to know where to look for our service. This is where **Service Publishing** comes into play.

You can think of service publishing as a directory to search for, or simply browse, services that want to be discovered for use. If you are old enough to know what the white pages are, that is a close analogy. Only, imagine that this directory not only helps you look up services, but manages security, offers advanced monitoring, and can even add a façade that allows API requests and responses to be translated on the fly. A façade could even call multiple APIs on the backend with a single composite API endpoint on the frontend. Many even support the mocking of APIs, so you can take your design-first approach, build an OpenAPI specification, then have one team implement a simple mock service while another implements the actual code base!

There are many tools and services out there that facilitate Service Publishing. You can even search the web for API indexes. If you are willing to spend some dollars, you can get access to some interesting information sources out there via published APIs.

The NASA public APIs are nice to play around with (`https://api.nasa.gov/`), although they do not currently expose an OpenAPI specification. No doubt this is in use somewhere behind the scenes. The main point is that you have somewhere to go, for a given organization, area of interest, or community group, where you can easily look for API services to consume.

Azure API Management

Azure API Management (or **Azure APIM**, as it is more commonly abbreviated) provides a vast amount of functionality to support any organization's API ecosystem. Specifically, we will focus on the use of API Service Publishing.

To demonstrate how we can easily publish services via APIM, we need to create an instance of the service in Azure.

> **Important note**
> The Azure API Management service can be costly to leave provision for an extended period. At the time of writing, you can create a Dev SKU instance of APIm for about $0.07/hour. You should be able to spend less than $1 to complete this exercise for yourself.

Let's create an APIm instance and publish our API definition:

1. Log in to the Azure portal (`https://portal.azure.com`) and search for **API management services**:

Figure 14.10 – Searching for API Management services

2. Click on the **Create** button and complete the various configuration pages:

    - Chose your subscription and resource group.

    - Give it a name and provide your contact details.

    - Set the pricing tier to **Developer**.

    - Leave all other settings as default.

    It can take 30–40 minutes for a new instance to provision.

3. Navigate to the newly created APIm resource and select the APIs menu blade. Then, choose **Create from definition**, using the **OpenAPI** format:

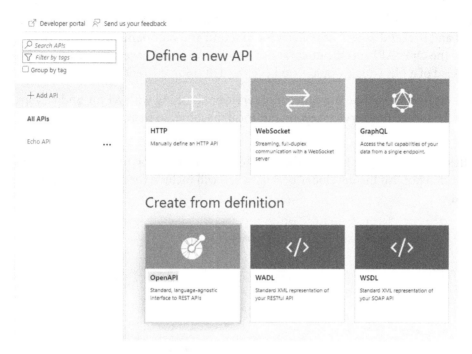

Figure 14.11 – Create from definition

4. Using the swagger.json file saved in the *Viewing the runtime OpenAPI specification* section, click on **Select a file**, upload it, and then click on **Create**:

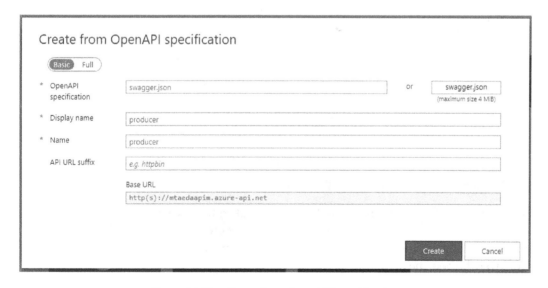

Figure 14.12 – Create from OpenAPI specification

At this point, you will see our API available for further configuring. We can configure a frontend and backend as well as policies for inbound and outbound processing. When we uploaded the OpenAPI specification for our service, we could have provided a runtime URL instead of a file, which would help keep the definition current.

Our focus is to demonstrate the concept of service publishing with APIm, and we have not actually pushed our service beyond our local machine, so there are no backends to configure.

Select the **Portal overview** blade in the **Developer portal** section of the APIm resource. Here, you can publish the developer portal, which will include the definition we just uploaded:

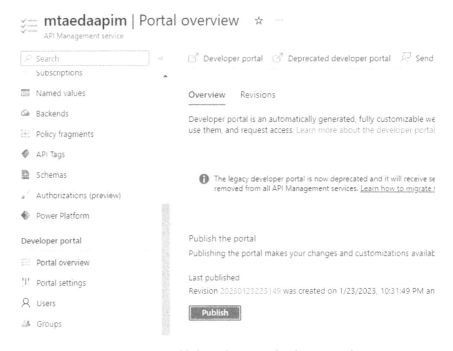

Figure 14.13 – Publishing the APIm developer portal

From this same blade, we can launch **Developer portal** using the button at the top.

By clicking on the **Developer portal** button, we are taken to a view that will be published to your organization's developers. As the APIm instance creator, you might that find it opens in an editable mode. You can *Ctrl + click* or *Command + click* the links to interactively browse the default site layout.

Click on the **Explore APIs** button, and then click on the **producer** service name. You will see another presentation of the content defined in the OpenAPI specification created earlier in this chapter:

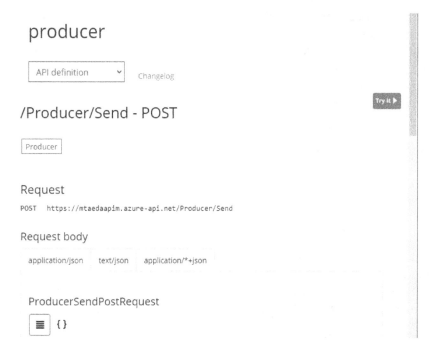

Figure 14.14 – The APIm presentation of our service, as published with the OpenAPI specification

That's it! We have made our API available for easy browsing across our organization. Typically, APIms will be centrally managed so that all service developments can be registered and controlled via the publishing service. The additional configurations to build out frontends, backends, mocks, authentication configurations, and policies all come into play for service publishing to truly work for your business. It is very common to see this publishing form part of a standard DevOps pipeline offered by the business to its developers for the easy and compliant onboarding of their services.

With the services defined, discovered, and published, what more could we possibly need to bring our API ecosystem to life? We should really think about service cataloging… but how does that differ from publishing? In this final section, we will understand the challenge and point to one growing solution example.

Service cataloging

With so many API services published, software supply chain management has unveiled an entirely new, and powerful, security risk.

Supply chain management is already a big challenge when dealing with static package management: verifying sources, attesting to a baseline trust of the maintainers, and staying current with security patches while being on the alert for malicious contributions!

Cloud security architects are popping up everywhere, and they have their work cut out for them. Managing basic authentication security patterns along with tackling limitless use of an uncountable number of APIs is daunting! External API services bring about serious security risks such as data exfiltration and malicious data inputs. But they also offer fast, innovative, and cost-effective ways to grow business capabilities. And you can't be slow… as your competitors are constantly trying to move faster.

Services such as Azure API Management offer a level of control. By ensuring your organizational developers can only access external services via APIm, we can at least provide a choke point from which to monitor and observe application dependencies on these external services.

This is a *lot* to keep track of, and API services are ever-changing, ever-moving, and ever-improving.

I want to make you aware of an open source project growing under the **Cloud Native Computing Foundation** (**CNCF**), which is trying to tackle this problem, **Ortelius**. The objective of **Ortelius** is to create a centralized supply chain evidence store that helps tackle the problem of continuously evolving dependencies on API services.

Ortelius allows API developers to register their services and track their consumers. In doing so, it is possible to create a dependency map across applications and the API services used.

Looking at API services and consumers is just one element of Ortelius' secret sauce for supply chain management. It also digests lots of other dependency and version information such as licensing, CVEs, source container registries, key values, and more. Applying this consistently across all your organizational assets, you can ask that famous question from 2021… "*Where am I using Log4j?*" Not only will you get a better answer than "*everywhere*," but you can also ask "*What is dependent upon an API service that uses Log4j?*" so that you can get a real idea of what your outages might look like as you patch that brutal **Log4Shell** vulnerability. Spoiler alert: the answer is "*everywhere*." But outside of an extreme example like this, the potential power of Ortelius to secure your business capabilities, while leaving your developers to do what they do best, should be an easy train to jump on.

Who knows whether Ortelius will become the future, or shape the future, as Worknick did? Either way, the problem it is addressing is vital to the continued growth of hyperscale, highly distributed systems with event-driven architecture.

Summary

In this final chapter, we've accepted that using event-driven architecture will result in the large adoption of microservice development. That brought about challenges of making services discoverable, well-defined, and published. We explored how we can document our API services with a common standard, utilize service discovery for fast-moving service instances, and service publishing to share our services with others. Finally, we looked at the security challenges supporting such a large, fast-paced ecosystem and tooling that is growing to tackle this challenge head-on.

Questions

1. What is the OpenAPI specification?

2. What is considered the best approach for creating an OpenAPI specification according to the creator?

3. What is the difference between SwaggerGen and SwaggerUI middleware?

4. How is Service Discovery better suited to microservices than traditional load balancing?

5. How can using a design-first approach to OpenAPI design and publishing it to an API management service help with development timelines?

Further reading

- *OpenAPI Initiative Home* – by The Linux Foundation, available at `https://www.openapis.org/`

- *OpenAPI Initiative Best Practices* – by The Linux Foundation, available at `https://oai.github.io/Documentation/best-practices.html`

- *Worknik Developer Documentation* – by Wordnik, available at `https://developer.wordnik.com/docs`

- *ASP.NET Core web API documentation with Swagger / OpenAPI* – by Microsoft, available at `https://learn.microsoft.com/en-us/aspnet/core/tutorials/web-api-help-pages-using-swagger?view=aspnetcore-7.0`

- *Get started with Swashbuckle and ASP.NET Core* – by Microsoft, available at `https://learn.microsoft.com/en-us/aspnet/core/tutorials/getting-started-with-swashbuckle?view=aspnetcore-7.0&tabs=visual-studio`

- *What is Service Discovery?* – by HashiCorp, available at `https://developer.hashicorp.com/consul/docs/concepts/service-discovery`

- *Consul on Docker Example Configuration* – by HashiCorp, available at `https://github.com/hashicorp/learn-consul-docker/tree/main/datacenter-deploy-service-discovery`

- *Raft – Understandable Distributed Consensus* – by Ben Johnson, available at `http://thesecretlivesofdata.com/raft/`

- *Tutorial: Import and publish your first API* – by Microsoft Learn, available at `https://learn.microsoft.com/en-us/azure/api-management/import-and-publish`

- *Versioning – Ortelius Secret Sauce* – by Ortelius, available at `https://docs.ortelius.io/guides/userguide/introduction/#versioning---ortelius-secret-sauce`

Assessments

Chapter 1, The Sample Application

1. Understanding the business perspective helps to tie real-world situations to conceptual designs as well as understand how behaviors of business objects can impact the need for resilient, scalable services to support them.

2. It's possible to theorize that there is a use case for a `WorkOrders` domain; however, the concept of `WorkOrder` may be better served as a supporting entity within the `Maintenance` domain.

3. Every effort was taken to ensure no erroneous or excess information was included in the aggregates. Personal preference or strict adherence to prescriptive DDD patterns may cause that to shift.

4. A traditional "*relational*" database tends to allow all forms of data manipulation, from inserts to updates to deletes while also structuring the data. NoSQL options allow for unstructured, heterogeneous data to be stored in the same logical data store. Using event sourcing strongly encourages the use of an append-only transaction log of events with a separate mechanism for querying the state of an object impacted by events.

5. Separating read and write operations gives the benefit of independent scalability of the services required. If one domain produces a lot of information and messages, it may make sense to increase the available resources for that write service to avoid unwanted performance issues. The same holds true for information that is commonly or heavily queried.

6. One advantage of separating the solutions is the ability to develop, test, build, and deploy those domain components independent of any other components in the application. This allows for more focused development of required features, more targeted testing, and a lack of dependence on the rest of the application when it comes to building and releasing that component. Segregating the solutions can introduce more complexity, as each solution would require its own pipelines and could also have different technologies used to implement core features such as read and write operations.

7. You can use any OpenID Connect authentication provider to secure calls to application components, in addition to JWT tokens, custom username/password stores, or certificates.

8. Using a standard schema for all events provides the advantage of consistency as well as the portability of those events to different analysis engines. It can add to the complexity of the solution as well, though, as writing and querying events from a particular topic may be more straightforward without the additional schema requirements. Your answer to this question will likely be more of a subjective one, based on experiences you have had.

9. The domain service projects can be deployed to many different targets aside from Docker or Kubernetes. Azure App Services, Windows services running on VMs, Linux-based services managed by systemd, and Azure API applications managed by an API management layer are all possible deployment and hosting targets.

Chapter 2, The Producer-Consumer Pattern

1. Yes; if producers were the only piece in existence, events could be written but never used. If consumers were the only piece in existence, no events would be produced for the consumers to consume.

2. There can be several obstacles around the adoption of the producer-consumer pattern and event-driven architecture. Moving from traditional client-server applications to decoupled event-driven components can take time, especially for those who have coded in a particular methodology for years. Existing application code can also hinder adoption as it might seem overwhelming to think of how much would have to change to embrace this new architecture. Not everyone learns at the same pace, so having realistic expectations about adoption can only help drive its success.

3. Minimal APIs are useful because they allow you to define routes and other functionality directly in the `Program.cs` file, which can help reduce the need to comb through source files for controllers or other code. For those coming from Node.js or Python backgrounds, it can feel familiar in that similar formats can be seen in those languages. It can have drawbacks such as having a great deal of code in one file, which may make searching for things more difficult if the file is ostensibly large. Additionally, it can be confusing for those who are used to API projects having many files for each controller, handler, ViewModel, or other object type.

4. The library being used is `Confluent.Kafka`, as supplied by Confluent. Confluent is a company that offers both managed and self-hosted options for Kafka with different platform features built-in.

5. The `BackgroundService` class is the base class used for those services. It can be found in the `Microsoft.Extensions.Hosting` package.

6. Yes; configuration files can be used to store XML or JSON representations of more complex objects. Using appropriate field names will ensure that binding a section of the configuration to a complex object is successful.

7. The usage of partitions is entirely dependent on the needs of the domain and the producers and consumers for that domain. In some cases, using multiple partitions can help stabilize throughput. In other cases, having one partition might make sense due to lower volumes or smaller record sizes.

8. A stream is a collection of records from a topic. They exist where a table is built from a stream and uses functions to enrich or alter the record data for ingestion elsewhere.

Chapter 3, Message Brokers

1. Queue-based, cache-based, and stream-based.

2. While AMQP is a popular messaging protocol, there are plenty of applications that use HTTP, TCP, MQTT, or other protocols to send and receive messages.

3. Yes. Whether through a traditional request-response or more modern mechanisms such as WebSockets, HTTP is perfectly viable as a transport mechanism.

4. This will depend on your overall application design choices and what works best for you and your team. Each approach has benefits and drawbacks.

5. At-most-once is the most likely to be susceptible to data loss.

6. Yes – so long as one functioning broker exists, a Kafka cluster can function.

7. Zookeeper tends to the different brokers within the cluster by providing configuration data, determining the active broker through leader election, and more.

8. No – topics can be created from the command line, using a GUI interface such as Kafdrop, or even programmatically via code when you're using an SDK to connect to a Kafka instance.

Chapter 4, Domain Model and Asynchronous Events

1. The maintenance domain.

2. The primary reason for using the mediator pattern is to decouple the producing and consuming code, allowing for a specific object to manage communications and reducing dependencies that could couple those consumers and producers together. The sample application uses the `INotification` and `INotificationHandler<T>` interfaces to differentiate event handlers within the domains and allow for known actions to be taken when an event handler is invoked.

3. No – asynchronous programming, as seen with the async/await pattern in C#, does not automatically create new threads when an async method is called. New background threads can be spun up as a result of an async method when using `Task.Run()` or `Task.Factory.StartNew()` to create a new task.

Chapter 5, Containerization and Local Environment Setup

1. Kubernetes, generally speaking, is a much more robust orchestration platform than Docker Compose. It also carries with it the burden of operational support and resource knowledge when building or scaling out a cluster. Docker Compose is a more streamlined approach to working with several different containers that comprise a logical development ecosystem but does not give you the array of resiliency, scalability, or reliability options that Kubernetes does.

2. While one could argue that all of the fundamentals mentioned could be beneficial, the portability and reusability aspects tend to shine through a bit brighter. Having a portable solution allows more people to work on a specific domain solution, and reusability allows for less churn when trying to set up stable code to test against.

3. The `depends_on` directive allows you to specify one or more services within the Docker Compose file that must be started before a specific service is started. Examples include starting Zookeeper before starting Kafka broker nodes or SQL Server before starting up the query API service.

4. Aside from platform constraints for memory and CPU (described in the GitHub documentation), the biggest constraint is that Visual Studio Code is the only supported IDE for use with Codespaces.

5. The main difference is that the Kafka and database services reference a pre-built image, whereas the domain services will build a local image from the source code prior to starting up, assuming any changes have been made to the service code.

Chapter 6, Localized Testing and Debugging of Microservices

1. The Dockerfile builds code in Release mode on an SDK image, publishes the files, and then copies them to a runtime-only image.

2. Visual Studio runs a container that is ready to build and execute code at runtime (instead of image build time). This container also has a remote debugger installed and configured, and a volume mapped to the local source code.

3. The Consumer will continue to read events from the Kafka cluster and process them as normal. Without the Producer, no more events will be written to the Kafka cluster.

Chapter 7, Microservice Observability

1. Kubernetes will continue to check the liveness endpoint. If it fails for longer than the configured tolerance, the Pod will be terminated. If the Pod is part of a deployment, it will attempt to create a new instance of the Pod.

2. Kubernetes will continue to check the readiness endpoint. If it fails for longer than the configured tolerance, Kubernetes will stop sending any requests to the Pod until the readiness endpoint succeeds again. In a deployment with multiple replicas, other Pods will share the incoming requests so long as their readiness endpoints succeed.

3. A correlation identifier will be consistently reported across the event logs related to a specific triggering action.

4. A causation identifier will be reported for each event, indicating the prior event that triggered the current event. This is a tree-like hierarchy, as one event could trigger multiple next events – especially using the publisher-subscriber pattern.

Chapter 8, CI/CD Pipelines and Integrated Testing

1. Not necessarily. Depending on the needs of the component itself, you may only have to implement one or two of the CI/CD patterns discussed. It is good practice to at least ensure that you are separating environments and using an artifact repository.

2. Normally, feature flags are seen as beneficial when working with an application that is driven by its user interface – that is, there is a primary user interface upon which the user experience is intended to be based. If your application is a set of microservices, for example, using feature flags may not be as relevant unless you have strict requirements to keep newer services relegated to only a specific subset of consumers.

3. While approval gates are normally the first item to come to mind, another frequent gate is that of change management. As an example, organizations using ticketing systems such as ServiceNow to manage how they raise, approve, and execute change requests can see substantial improvements in life cycle time for changes if integrations are enabled with the CI/CD pipelines. One method for doing this would be to allow non-production code changes to be considered less risky and automatically approved as changes, thus enabling CD pipelines to push new code without approval overhead. The review and approval of production changes would be a necessary gate, but once the item has been approved in ServiceNow, it could prompt the CD pipeline to then execute and deploy new code to production based on that approval.

4. This is widely open to interpretation. You will want to select an approach to running your integration tests (one of the four mentioned earlier in this chapter), but implementing those in your pipelines can truly be a team decision. You may find that the organization you work for prefers to run these types of tests using a prescriptive approach, but how you orchestrate those is less important than ensuring the approach is adhered to.

Chapter 9, Fault Injection and Chaos Testing

1. A fault is something that occurs within a circuit, computer hardware, operating system, or other hosting mechanisms that cause a failure that cannot be recovered from easily. An exception is generally seen as a construct within software development where an error occurs during execution, which may or may not be expected.

2. A common fault that can be anticipated and handled is that of a regional service outage for a cloud-native service. For example, if there is an outage affecting App Services in the US East region of Azure, this could be handled by keeping a secondary copy of those App Services in another region and configuring them to be available in the event of such an outage.

3. Fault injection can be seen as an isolated discipline. However, in the software world, it doesn't have to be. Using fault injection as a method to carry out chaos experiments is one that is commonly used to illustrate the interconnected role that fault injection can play. In hardware-based scenarios, you may argue that fault injection is more of a standalone concept mainly because, if successful, it results in the complete malfunction of a product, which may not be recoverable.

4. Fault injection, as well as chaos testing, can take place in a variety of locations and environments, and by a variety of different means. Common locations are integration or QA environments, though some will run fault injection tests in production to gain insights. These tests can be run through pipeline automation, script runners, or even manually. The preferred approach, as with anything CI/CD-related, is to automate as much as possible.

5. The initial impact of a chaos experiment, or blast radius, should start small and increase in size until the experiment exposes an issue, or proves the application is tolerant of the experiment's conditions. You might start with a single record in a database or a single secret within a secret store, then expand to include more and more areas until the entire database, secret store, or both are impacted.

6. According to the list of fault providers supplied by Microsoft, the following types can be used as targets for a chaos experiment: Azure Cache for Redis, Domain Names (classic), virtual machines, virtual machine scale sets, Key Vault, Network Security Groups, Azure Kubernetes Service, and Cosmos DB.

Chapter 10, Modern Design Patterns for Scalability

1. No; virtual and physical networks are created with a finite amount of resources and cannot be autoscaled. Monitoring conditions related to network traffic can be a means of triggering an autoscaling event to better handle an influx of traffic.

2. CPU and memory tend to be the easiest resources to adjust and will generally be seen as the targets for monitoring usage or over-usage.

3. Out of the box, HPAs support evaluating thresholds on CPU and memory usage relative to the pod(s) assigned to the application. While other types of metrics can be leveraged, from monitoring other cluster components or even outside hosted systems, the recommendation is to stick with usage patterns with HPAs that are supported by default.

4. In the short term, yes. Longer term, even increasing the number of resources for the cluster nodes can become less maintainable and take away from your ability to separate workloads efficiently. Having a separation between system-critical pods in the default pool versus application-critical pods in an additional node pool is a good way to promote independently scalable workloads without worrying about consuming the resources needed by Kubernetes to operate the cluster.

5. This may be more of an opinion… technically, the use of metrics related to external systems is less of a best practice with the native HPA resources in a cluster. There are plenty of valid

upstream dependencies that could trigger an autoscaling event in the cluster to proactively help the application handle a surge in traffic. Handling events, after all, is the main theme of the system architecture for our application.

Chapter 11, Minimizing Data Loss

1. Immediate consistency/ACID

2. Eventual consistency/BASE

3. **Command query responsibility segregation (CQRS)**

4. Eventual consistency/BASE

5. At-most-once delivery

6. At-least-once delivery

7. Effectively-once delivery

Chapter 12, Service and Application Resiliency

1. Resiliency refers to an application or platform's ability to handle faults, exceptions, or other problems that may arise. Redundancy refers to the duplication of data or services (or both) to help protect against data loss due to network issues or in the event of a disaster. Reliability is a measure of confidence in a platform or application's ability to operate consistently.

2. Both fallback and circuit breaker patterns allow you to stop execution and return a value that can be used in lieu of the expected operation. Fallback patterns will not perform any additional logic to allow further operations to continue on the same target, whereas circuit breakers will pause execution for a set amount of time and reset once the threshold is met.

3. It allows you to retry operations that may fall victim to timeouts or other connectivity issues, either through graduated wait times or by coupling with other patterns to manage timeouts, circuit breakers, or other mechanisms.

4. Disaster recovery refers to a plan that supposes a data center has been taken offline and outlines a secondary site that would be able to be activated to restore application functionality. Business continuity deals with preserving the continuity of services, regardless of whether the outage is due to a temporary outage or a disaster with a data center loss.

5. It's possible to run the services using a lightweight Kubernetes cluster on a server at the station or to utilize storage accounts to house events instead of a service bus instance. There could be other potential solutions as well, depending on pricing and domain needs. This is by no means an exhaustive list of alternatives.

Chapter 13, Telemetry Capture and Integration

1. Sometimes, the platform used to capture telemetry is not at the developer's discretion, and another means of capturing and analyzing the information will be used. Common examples include the ELK stack and Splunk. Teams may also want tooling that is independent of a particular cloud but common across a specific cloud service, such as Kubernetes.

2. The `Activity` type is generally used during the course of method execution to capture information often found useful for tracing, while `Meter` objects are generally about capturing an occurrence or other frequencies over time related to actions within the code. How the information is measured is one of the primary drivers, along with the implementation model.

3. Yes, each of those examples listed will factor into the overall design and usage of any telemetry capture platform. Depending on what is chosen, the cost can vary from data storage to transactions processed to individual licenses to use the product. Constraints on platform capabilities should always be considered when choosing to use the platform in question. Performance can come into play when services are too chatty, rely on synchronous means of collecting information, or if the platform is self-maintained; the improper sizing of resources to host the platform can also impact performance.

Chapter 14, Observability Revisited

1. The OpenAPI specification is a standard format used to describe available REST API methods, including their detailed call signatures and return structure.

2. Design-first is the best practice for creating an OpenAPI specification.

3. SwaggerGen serves the OpenAPI specification for your service in JSON format. SwaggerUI renders that specification as an interactive web page for easier browsing and use.

4. Service discovery is designed to handle the fast-paced creation and removal of service instances across multiple environments.

5. Once APIs have been designed and published; they can be supported by mock implementations and shared across teams. This allows multiple teams to design and develop against the use of this intended API service even before it has been coded as a real functioning service.

Index

Other Books You May Enjoy

If you enjoyed this book, you may be interested in these other books by Packt:

Microservices Design Patterns in .NET

Trevoir Williams

ISBN: 9781804610305

- Use Domain-Driven Design principles in your microservice design
- Leverage patterns like event sourcing, database-per-service, and asynchronous communication
- Build resilient web services and mitigate failures and outages
- Ensure data consistency in distributed systems
- Leverage industry standard technology to design a robust distributed application
- Find out how to secure a microservices-designed application
- Use containers to handle lightweight microservice application deployment

Parallel Programming and Concurrency with C# 10 and .NET 6

Alvin Ashcraft

ISBN: 9781803243672

- Prevent deadlocks and race conditions with managed threading
- Update Windows app UIs without causing exceptions
- Explore best practices for introducing asynchronous constructs to existing code
- Avoid pitfalls when introducing parallelism to your code
- Implement the producer-consumer pattern with Dataflow blocks
- Enforce data sorting when processing data in parallel and safely merge data from multiple sources
- Use concurrent collections that help synchronize data across threads
- Debug an everyday parallel app with the Parallel Stacks and Parallel Tasks windows

Packt is searching for authors like you

If you're interested in becoming an author for Packt, please visit authors.packtpub.com and apply today. We have worked with thousands of developers and tech professionals, just like you, to help them share their insight with the global tech community. You can make a general application, apply for a specific hot topic that we are recruiting an author for, or submit your own idea.

Share Your Thoughts

Now you've finished *Implementing Event-driven Microservices Architecture in .NET 7*, we'd love to hear your thoughts! Scan the QR code below to go straight to the Amazon review page for this book and share your feedback or leave a review on the site that you purchased it from.

https://packt.link/r/1-803-23278-1

Your review is important to us and the tech community and will help us make sure we're delivering excellent quality content.

Download a free PDF copy of this book

Thanks for purchasing this book!

Do you like to read on the go but are unable to carry your print books everywhere? Is your eBook purchase not compatible with the device of your choice?

Don't worry, now with every Packt book you get a DRM-free PDF version of that book at no cost.

Read anywhere, any place, on any device. Search, copy, and paste code from your favorite technical books directly into your application.

The perks don't stop there, you can get exclusive access to discounts, newsletters, and great free content in your inbox daily

Follow these simple steps to get the benefits:

1. Scan the QR code or visit the link below

https://packt.link/free-ebook/9781803232782

2. Submit your proof of purchase
3. That's it! We'll send your free PDF and other benefits to your email directly

www.ingramcontent.com/pod-product-compliance
Lightning Source LLC
Chambersburg PA
CBHW062104050326
40690CB00016B/3199